THE ANTHROPIC PRINCIPLE

A Universe Built for Man

Anthony Walsh, Ph.D.
Boise State University

Series in Philosophy of Religion

VERNON PRESS

Copyright © 2023 VernonPress, an imprint of Vernon Art and Science Inc, on behalf of the author.

All rights reserved. No part of this publication may be reproduced, stored in a retrieval system, or transmitted in any form or by any means, electronic, mechanical, photocopying, recording, or otherwise, without the prior permission of Vernon Art and Science Inc.
www.vernonpress.com

In the Americas:
Vernon Press
1000 N West Street, Suite 1200,
Wilmington, Delaware 19801
United States

In the rest of the world:
Vernon Press
C/Sancti Espiritu 17,
Malaga, 29006
Spain

Series in Philosophy of Religion

Library of Congress Control Number: 2022945869

ISBN: 978-1-64889-585-2

Also available: 978-1-64889-524-1 [Hardback]; 978-1-64889-551-7 [PDF, E-Book]

Product and company names mentioned in this work are the trademarks of their respective owners. While every care has been taken in preparing this work, neither the authors nor Vernon Art and Science Inc. may be held responsible for any loss or damage caused or alleged to be caused directly or indirectly by the information contained in it.

Every effort has been made to trace all copyright holders, but if any have been inadvertently overlooked the publisher will be pleased to include any necessary credits in any subsequent reprint or edition.

Cover design by Vernon Press using elements images designed by Rochak Shukla on Freepik.

Table of Contents

List of Figures — v

Preface — vii

CHAPTER ONE
The Copernican Principle or the Anthropic Principle? — 1

CHAPTER TWO
God or Science? The Conflict that Never Was — 11

CHAPTER THREE
The Miracle of Mathematics — 21

CHAPTER FOUR
Finding God in the Micro World: The Standard Model of Particle Physics — 31

CHAPTER FIVE
Finding God in the Macro World: The Cosmos — 41

CHAPTER SIX
Our Cosmic Neighborhood — 51

CHAPTER SEVEN
Our Very Special Earthly Home — 61

CHAPTER EIGHT
Avoiding Anthropic Fine-Tuning with the Multiverse — 71

CHAPTER NINE
The Molecules of Life — 81

CHAPTER TEN
The Queen of all Scientific Problems: The Origin of Life — 91

CHAPTER ELEVEN
DNA: God's Book of Life — 103

CHAPTER TWELVE
Evolution by Natural Selection: Micro and Macro   113

CHAPTER THIRTEEN
I Am Fearfully and Wonderfully Made   125

CHAPTER FOURTEEN
Mind, Consciousness, Language, and Free Will   137

References   149

Index   161

List of Figures

Figure 3.1.	The Golden Spiral and Rectangle Formed by the Fibonacci Sequence	27
Figure 4.1.	Standard Model of Elementary Particles	33
Figure 5.1.	The Timeline for the Expansion of Space from the Big Bang	46
Figure 6.1.	The Electromagnetic Spectrum	56
Figure 7.1.	Habitable Zones Around Different Types of Stars	61
Figure 8.1.	The Triple-Alpha Process	86
Figure 11.1.	The Making of a Protein	107
Figure 13.1.	The Neuron and its Parts	133

Preface

The Copernican Principle has been used by some to state that humankind is an insignificant assemblage of chemical scum living on an accidental planet in a suburb of a purposeless universe. This scurrilous principle has been questioned by many prominent scientists, including Nobel laureate physicists, which has led physicists to propose the Anthropic Principle. This principle posits a purposeful link between the structure of the universe and the existence of humankind and its specialness. The numerous features of the universe are so freakishly fine-tuned for the existence of intelligent life that physicists are beginning to come to grips with the notion that our universe is profoundly purposeful and that there is a powerful and incredibly intelligent Mind behind it all.

The four primary versions of the principle are the: Weak Anthropic Principle (WAP), the Strong Anthropic Principle (SAP), the Final Anthropic Principle (FAP), and the Participatory Anthropic Principle (PAP). WAP simply says that our location in the universe is privileged because it is compatible with our existence, and SAP says that the universe had to result in the creation of intelligent life at some point. This implies purpose and deliberate design behind the universe. FAP says that once intelligent life comes into existence it will never die. The idea that God created the universe as a home for humans is unattractive to atheists, but a number of scientists have been forced to God by anthropic fine-tuning. PAP proposes that observers are necessary to bring the universe into existence. This is consistent with the standard interpretation of quantum mechanics. There is no quantum reality until an observer exists to witness wave collapse. PAP's idea is that an intelligent observer imparts reality to the universe, but if the pre-human universe was "observed" into being, the only candidate for the job must be the Ultimate Observer—God.

Chapter two examines the allegation that science and Christian theism are in conflict. Although it is true some *scientists* are at war with theism, *science* itself is not. The spirit of science grew out of the Christian belief in a rational and orderly God who created us in His image, and many of the advances in early science were made by men of God. Scientists readily acknowledge that the big questions of meaning are outside of their purview, so if we are to find answers to ultimate questions, we need both science and God. Science seeks answers to *how* God created the universe; theology searches for *why* He did. Science works within a materialist/naturalist framework, and this is amply justified. The problem comes when we jump from a working assumption to the assumption that there is nothing beyond the realm of the material/natural.

Mathematics is the subject of chapter three. Early scientists such as Copernicus, Galileo, Kepler, and Newton knew that the universe was capable of mathematical description because a rational God fashioned it in a rational way, and later Nobel laureates such as Roger Penrose and Paul Dirac have concurred. Mathematical truths represent the real world in abstract symbols and have been amazingly successful in doing so. I give examples of this, including the probability boundary—the point at which something improbable becomes impossible. I also look at the mysterious golden ratio, Fibonacci numbers, and the Fibonacci cascade that describe so many diverse and unrelated phenomena.

Chapter four addresses the Standard Model of particle physics, including the Higgs boson, whimsically nicknamed the "God Particle." Many physicists regard the Standard Model as highly "unnatural" because of the large number of parameters that are balanced on a razor's edge such that changing any of the values and we would have a universe without atoms. "Naturalness" is the prohibition of anthropic fine-tuning. Since we have a universe with fine-tuned to multiple trillions of degrees more than the Standard Model predicts, the unnatural universe must be supernatural. The four fundamental forces of nature—gravity, electromagnetism, and strong and weak nuclear forces—are addressed in terms of their remarkable fine-tuning.

We move from the micro to the macro in chapter five with the Big Bang. Before the Big Bang, almost all scientists believed that the universe was static and existed eternally. There was fierce opposition to the Big Bang theory because it is reminiscent of Genesis's creation from nothing. It was not until the evidence for an expanding universe was buttressed by the discovery of the cosmic microwave background radiation was discovered that almost all scientists accepted the Big Bang. I look at Penrose's calculations of the utter impossibility of getting the whole show on the road given that it had to begin with the lowest possible entropy level, the geography of the universe, and the "unnaturalness" of the cosmological constant.

Chapter six looks at our cosmic neighborhood—the Milky Way, other types of galaxies, and the process of nuclear fusion in stellar nucleosynthesis and supernovae. We are in the galaxy's "sweet spot:" referred to as the galactic habitable zone. For a variety of reasons, this is the only area suitable for life. The Sun's perhaps unique properties are examined and how they make our planet suitable for life such as its stable luminosity and its vital role in stabilizing the tides. I then look at other anthropic considerations relating to the Sun, and why it has been used as a metaphor for God.

We look at our prize patch of cosmic real estate called the Earth in chapter seven. The Earth is located in the solar system's Circumstellar Habitable Zone, which is a band of space around the Sun that is hospitable to life. The Earth's location, orbital eccentricity, mass, magnetic shield, plate tectonics, and ozone

Preface ix

layer, among many other things, contribute to its habitability. The wonders of photosynthesis, the process by which plants obtain their food and animals their oxygen, are then briefly discussed, along with the role of volcanoes. This is followed by a discussion of the Moon, Jupiter, and Saturn in making our planet safe and habitable.

The molecules of life are the topics of chapter eight, beginning with water. Despite its simple structure, water is the strangest liquid on the planet because it bends the rules of chemistry, but life is impossible without it. The same can be said of carbon because it forms the backbone of millions of organic compounds. Carbon is forged in the stars, but scientists were at a loss as to how until Fred Hoyle made his anthropic prediction that it was done via a "triple alpha" process. I then take a broader look at the marvel of photosynthesis, how it proceeds, and what it does for us in providing food and oxygen. I finish by looking at nitrogen, the most abundant element in the Earth's atmosphere and nature's chief fertilizer.

Many scientists hate the unnatural anthropic fine-tuning we observe and have turned to speculations about a multiverse to avoid it; posit enough universes and one can beat the odds of finding one with its parameters fine-tuned to such an incomprehensible degree as ours. Chapter nine examines the various multiverse models based on M theory, the mathematic basis for the multiverse. Multiverse proponents know that the theory cannot, even in principle, be empirically tested and argue for a relaxation of the way a theory should be accepted. Even if the multiverse turns out to be right, it does not mean that it would exclude God. If God is capable of creating one universe, He is capable of making trillions, so the choice is hardly God or the multiverse.

Chapter ten examines abiogenesis, the hypothetical process by which chemical evolution became biological evolution. The leap from non-living matter to living matter would require a set of random lifeless molecules to arrange themselves in specific and complex ways to gain both a metabolic and a reproductive capacity, the systems that define life. I examine the two major hypotheses of abiogenesis; the RNA world, and the metabolism first hypotheses, before looking at a newer idea that abstract information came first. Recognizing the difficulties for a naturalistic emergence of life on Earth, some have turned to the multiverse and to the notion of panspermia to get around it.

The cell, genome, and DNA are the topics addressed in chapter eleven. These marvels of nanotechnology are God's construction manuals that provide the information needed to build the proteins that build us. I briefly look at the structure and functions of various parts of the cell, the process of going from the information content of DNA to proteins, the intricacies of protein folding, and the work of the Encyclopedia of DNA Elements (ENCODE) consortium that has found function in thousands of stretches of DNA that used to be dismissed as "junk DNA."

Darwin's evolution by natural selection is the topic of chapter twelve. While not disavowing it, there are many difficulties involved, including the huge "waiting time" involved even for the mutation of even a functioning enzyme to become a different one. Micro- versus macro-evolution, chance and necessity, the Cambrian explosion, punctuated equilibrium, and the tree of life are explored. After struggling with evolutionary theory for decades, I have allied myself with theistic evolution (TE). TE believes that God created all living things using the process of evolution in ways that conform to secular scientific accounts, but denies that evolution is undirected and purposeless. The ideas of major scientists and theologians, including Augustine and Thomas Aquinas, on TE are discussed.

Chapter thirteen discusses the human body, the most complex thing in the universe, and the brain, the most complex part of the body. We begin with the profound mystery of why the zygote exists because sexual reproduction seems highly unlikely given the "simplicity" of asexual reproduction. Mitosis and meiosis, the innate and acquired immune systems, the cardiovascular system, and the eye are discussed. This is followed by a discussion of the brain; God's magnum opus. We look at the various parts of the brain and their functioning, including the process of synaptogenesis, the process by which the brain incorporates environmental experiences into the brain's neurocircuitry, and why love is so important in this.

The final chapter addresses attributes of humans that most distinguish them from other animals—mind, consciousness, language, and free will. Materialists imagine that the mind is just the brain at work, but it is just as easy to imagine the opposite. Consciousness—being aware that we are aware—and communication via language has enabled humans to actively make their environments rather than merely adapting to them. But are we really only responding to our genetic makeup and environmental exigencies when we make decisions, or do we have free will? The answer to this depends on how we define free will and determinism. Neither free will nor determinism alone is sufficient to explain human behavior; we need both to do so.

Chapter One

The Copernican Principle or the Anthropic Principle?

The Copernican Principle

The Bible tells us that humans are made in the image of God, which means that humans hold a very privileged status in the universe. Non-believers may tell you that this is inexcusably arrogant and that Christians should learn some humility. Being made in the image of God does not mean humans are corporeal representatives of the imageless God; this reduces God to human proportions. So, what does it mean? Theologians have been arguing this for centuries. Is the meaning found in its relational sense; in the human capacity for a relationship with God and with one another? Is it the covenant we have with God, or perhaps it is something we *do* rather than what we are or what we have? In Genesis, God was creating and delegated us the authority to do the same: "Be fruitful, multiply, fill the earth, and subdue it." Thus, being made in the "likeness" of God, means doing on a human scale what He did—create, love, behave morally, justly, and mercifully. If we can do this, we are special.

Stephen Hawking says no! and that not only are we not special, we are downright insignificant: "just a chemical scum on a moderate-sized planet, orbiting around a very average star in the outer suburb of one among a hundred billion galaxies" (in Kahane, 2014, p. 745). We might call this the misanthropic principle. Others want to demote humans too, if not as low as chemical scum, at least to just another animal separated from the rest only by arrogance. It is often said that the image we have of ourselves contains our destiny. In *The Myth of Sisyphus*, atheist philosopher Albert Camus explored the absurdity of a Godless life shorn of meaning. The opening lines of his book are: "There is but one truly serious philosophical problem, and that is suicide. Judging whether life is or is not worth living amounts to answering the fundamental question of philosophy" (Camus, 1955, p. 3). If there is no purpose in life other than to indulge in our natural appetites promiscuously, as Camus advised, and if we believe that we are just "chemical scum" living on a paltry piece of space rock with no sense of ultimate meaning, we may indeed feel that life is not worth living.

The notion of human mediocrity asserts that intelligent life is likely duplicated on billions of other planets in the universe. It is more formally known as the Copernican Principle, named after fifteenth-century Polish mathematician and astronomer Nicolaus Copernicus. Copernicus had absolutely nothing to do with the principle applied to his name and would have been saddened and perplexed to have his name sullied by association with it because he was a God-inspired man. The phrase was introduced into the philosophical lexicon in 1952 by mathematician Hermann Bondi in support of his now-discredited steady-state cosmology (Bondi, 1952). The steady-state theory was formulated in opposition to the Big Bang theory, which asserted that the universe was created at a specific point in time, which was philosophically unacceptable to atheists. Bondi, and other supporters of the steady-state theory, accepted the Aristotelian notion that the universe had no beginning; that is, the universe is past-eternal.

What Copernicus actually did was to advance the heliocentric ("Sun-centered") view of our solar system opposing the then generally accepted geocentric model. The geocentric model is most associated with the second-century Egyptian astronomer Claudius Ptolemy. Ptolemy's model declared that the Earth is literally at the geometric center of the then-known universe and the Sun, stars, and planets revolved around it. Copernicus' model removed Earth, and thus its inhabitants, from the center of the universe. Of course, the Earth was never at the "center" of the solar system, much less the universe, because the universe has no center. Using heliocentrism to demote humans to mediocrity confuses place with value. The heliocentric model simply shows that the Earth is not in a central *location* in the solar system; it says nothing at all about the worth of its inhabitants.

The debate surrounding the two models reached its height with the 1633 conflict between Galileo Galilei and the Catholic Church. Unfortunately, the Galileo story has come down to us as a narrative of the church denying science, but the truth is that Galileo's trial lasted only one day and resulted in his house arrest for life. He had previously gained the blessing of the Church to investigate both models, but the sixteenth-century Protestant Reformation moved the Church to tighten its grip on authority in this time of great religious instability. It is little known that Galileo considered scientists, not the Church, to be the chief opponents of the heliocentric theory advanced by Copernicus 90 years earlier. By this time, it was thought that the telescope could settle the argument between the two models, but the telescope was not welcomed by all. As Galileo wrote to his friend, German mathematician, and astronomer, Johann Kepler "What do you have to say about the principal philosophers of this academy who are filled with the stubbornness of an asp and do not want to look at either the planets, the moon or the telescope, even though I have freely and deliberately offered them the opportunity a thousand times?" (in Duck and Duck, 2014, p. 32).

In Galileo's time, most natural philosophers (scientists) subscribed to the geocentric model while the Jesuits, a religious order with a solid scientific reputation, questioned it. Cesare Cremonini, a colleague of Galileo's, accepted the challenge and looked through the telescope, but complained that it gave him a headache and would hear no more about it. For Cremonini to have endorsed the evidence before his eyes would have called into question his life's work, for he was a scholar of Aristotelian philosophy, and Aristotle had advanced a geocentric model before Ptolemy. In Cremonini's defense, the heliocentric model was ridiculed almost universally by scientists as defying common sense. We can understand that ridicule because we would all subscribe to the geocentric model if all we had to go by were our unaided sensory observations. The geocentric theory comports with our sensory experiences in a way that the heliocentric theory does not. After all, we don't feel Earth moving as it spins on its axis at just over 1,000 miles per hour while hurtling through space in orbit around the Sun at about 67,000 miles per hour, and we see the Sun rise in the east, move across the sky, and then set in the west.

The Anthropic Principle

Unfortunately for the Copernican Principle, scientific advances since Bondi coined the phrase have shown that the universe is not only special but is a little too special for atheists. Nobel laureate physicist Robert Millikan once wrote that the more we investigate the wonders of existence the more "we recognize the existence of a Something, a Power, a Being in whom and because of whom we live and move and have our being—a Creator by whatever name we may call Him" (in Walsh, 2020, p. 23). As we discovered more and more about our universe, many physicists began to ponder if its parameters, so exquisitely fine-tuned, exist for the benefit of humanity. The possible values that these physical parameters could have taken make the probability that they have the values that they do astronomically small. Scientists who give much thought to these things often come to the same conclusion as Millikan: there is a powerful and incredibly intelligent Mind behind it all.

The pairing of scientific observations with philosophical reasoning evolved into an idea called the Anthropic ("human-centered") Principle, which arrived almost simultaneously with the Copernican Principle in the mid-twentieth century. The phrase was coined by physicist Brandon Carter and is the polar opposite of the Copernican Principle's notion that we are accidental creatures in an accidental universe. John Wheeler, one of the greatest physicists of the twentieth century, contrasts the Copernican and Anthropic principles: "Is man an unimportant bit of dust on an unimportant planet in an unimportant galaxy somewhere in the vastness of space? No! The necessity to produce life lies at the center of the universe's whole machinery and design" (in Ofulla, 2013, p. 139). Anything designed, of course, requires a designer.

The Anthropic Principle is seen as controversial because statements such as Millikan's and Wheeler's imply a purposeful link between the structure of the universe and the existence of mankind and human specialness. The numerous features of the universe that are so freakishly fine-tuned for the existence of intelligent life that many physicists are beginning to come to grips with the notion that our universe is profoundly "unnatural." As physicist Nima Arkani-Hamed declared in a talk at Columbia University (2013): "The universe is inevitable," and at the same time "The universe is impossible." How can something be both inevitable and impossible? There is no other explanation for this apparent contradiction than the Anthropic Principle.

There are a variety of versions of the principle; the first being the Weak Anthropic Principle (WAP). The WAP is defined by Carter as: "we must be prepared to take account of the fact that our location in the universe is necessarily privileged to the extent of being compatible with our existence as observers" (1974, p. 293). Many dismiss WAP by pointing out that it is not at all surprising that we see this compatibility since if the universe were not so we wouldn't be here to discuss it. This is an obvious but question-begging response because it does not inform us of *why* we are here to discuss it when it is overwhelmingly more likely that we should not be. Philosopher John Leslie rebutted this objection to WAP with his "firing squad" analogy. He asks us to imagine a condemned man facing a firing squad of 100 marksmen. The order to fire is given, the shots ring out, but the man walks away unscathed. It is possible that one marksman missed, but surely it is impossible that all did. It is not sensible to say that it is not at all surprising, since if they had not all missed, the condemned man would not be alive to walk away and tell the tale. Why he walked away demands an explanation. It is more sensible to conclude that something intentional was afoot; that is, the firing squad was designed such that the condemned man should go on living (Leslie, 1989). We can apply the same reasoning to the universe—there is something intentional afoot.

It is difficult at first blush to see how physicists would find such an apparent truism as WAP useful, but physicist Frank Tipler observes: "But the Weak Anthropic Principle is not trivial, for it leads to unexpected relationships between observed quantities that appear to be unrelated!" (1988, p. 28). Among other eminent physicists who invoke WAP is physicist Andrei Linde, who opines: "Those who dislike anthropic principles are simply in denial…One may hate the Anthropic Principle or love it, but I bet that eventually, everyone is going to use it" (in Susskind, 2005, p. 353). It is a short step from the Anthropic Principle to a design argument, as physicist Josip Planinić points out: "The anthropic principle, or the fine-tuned universe argument, can also be put forward as a design argument...It seems that the universe is arranged (tuned) exclusively to be agreeable to man. This thought on the notion of purposefulness implies the existence of a Creator of the universe" (2010, p. 47).

Carter later added the Strong Anthropic Principle (SAP), which says: "The universe (and thus the fundamental parameters on which it depends) must be such as to admit the creation of observers within it at some stage" (1974, p. 294). SAP takes note of the many astonishing coincidences between different branches of physics that work together against mind-boggling odds to make intelligent life possible. Carter's statement strongly implies purpose and deliberate design behind the universe.

As Freeman Dyson notes: "As we look out into the Universe and identify the many accidents of physics and astronomy that have worked together to our benefit, it almost seems as if the Universe must in some sense have known that we were coming" (1979, p. 250). Atheists recoil at the notion of a purposeful universe, but no less a mind than Albert Einstein believed in one: "The religious inclination lies in the dim consciousness that dwells in humans that all nature, including the humans in it, is in no way an accidental game, but a work of lawfulness that there is a fundamental cause of all existence" (in Isaacson, 2007, p. 20). There is no other reasonable explanation of why the universe had to "admit the creation of observers;" an argument for an endless trail of monstrously improbable "fortuitous coincidences" just won't wash.

Barrow and Tipler then proposed the Final Anthropic Principle (FAP), which says: "Intelligent information-processing must come into existence in the universe, and, once it comes into existence, it will never die out" (Barrow and Tipler, 1986, p. 23). The FAP is reminiscent of a basic tenet of Christian faith as set forth in John 3:16: "For God so loved the world that He gave His only begotten Son, that whoever believes in Him shall not perish, but have eternal life." The idea that a Supreme Being created the universe as a home for intelligent life is most unattractive to the committed atheist. But atheist scientists have to bump into the Anthropic Principle in their work occasionally. When they do, they may ignore it, explore it, or attempt to explain it away. As physicist Heinz Pagels puts it: "Faced with questions that do not neatly fit into the framework of science, they are loath to resort to religious explanation; yet their curiosity will not let them leave matters unaddressed. Hence, the anthropic principle. It is the closest that some atheists can get to God" (1985, p. 38). In fact, a number of atheists have been forced to God by contemplating the many incidences of anthropic fine-tuning. Frank Tipler is one of them who completely changed his worldview:

> When I began my career as a cosmologist some twenty years ago, I was a convinced atheist. I never in my wildest dreams imagined that one day

I would be writing a book purporting to show that the central claims of Judeo-Christian theology are in fact true, that these claims are straightforward deductions of the laws of physics as we now understand them. I have been forced into these conclusions by the inexorable logic of my own special branch of physics. (1994, p. i)

The Participatory Anthropic Principle (PAP) was proposed by physicist John Wheeler. PAP posits that observers are necessary to bring the universe into existence. This sounds weird, but it is consistent with the standard interpretation of quantum mechanics that governs the behavior of subatomic particles. In the strange world of quantum mechanics, all subatomic matter is in a state of wave-like "superposition;" that is, in all possible states at once. The uncertainty principle states that it is impossible to know both the position and momentum of say, an electron, at any point along its wave-like trajectory around the atom. When a wave function "collapses," it reduces to a single particle-like state with a definite location. There is no quantum reality until an intelligent observer exists to witness the collapse of the wave function. Thus, it is not too farfetched to argue that intelligence is necessary to make the universe real. Of course, the physical universe must first exist to provide the necessary elements for life, but the idea behind PAP is that the intelligent observer then symbiotically imparts meaningful reality to the universe. If the pre-human universe is observed into being, the only candidate for the job must be the Ultimate Observer—God. Quantum physicist Robert Russell (2008a) makes this point in his theory of NIODA (non-interventionist objective divine action).

Scientific Explanation and Anthropic Reasoning

Scientific explanations are meant to provide an understanding of the natural world. It assumes that a set of claims are accurate if they are supported by evidence that other scientists can evaluate. Explanatory reasoning involves mixtures of deduction, induction, and abduction. Deduction is a "top-down" method that reasons from a premise that is self-evidently true to a conclusion that logically follows. Deduction reasoning is most evident in mathematics because mathematical inquiries always begin with self-evident (*a priori*) truths. Because mathematical thinking rests on propositions that are true by definition, it is idealized as the preferred path to the truth by a school of thought called rationalism. If $x = 2$ and $y = 3$, then $x.y = 6$ is absolute in all possible instances. Rationalists contend that the world can only be understood through the intellect because the senses allow us only to see it as *it appears*. They say that the world comes to us through the buzzing confusion of sense perceptions and must be filtered and organized by the intellect. It is true that our senses may deceive us, and that whatever we experience with them must be mentally interpreted. But the intellect also deceives even the greatest of minds. Nevertheless, deductive reasoning from self-evident truths has been taken as the ideal path to knowledge because it guarantees the truth of the conclusion given that it is already present in the premise ("All crimes are against the law.") and any denial of it is self-contradictory (if an act is not against the law, it cannot be a crime, even if we think it should be).

Outside mathematics deductive reasoning can let us down. Except in the most trivial of senses ("All mothers are females"), we have few major premises that are self-evidently true. Knowledge must be gained by observation and experiment by "bottom-up" reasoning from the specific to the general. This is known as induction. A "conclusion" in a deductive mode is a "hypothesis" in an inductive mode; an assertion to be tested experimentally. To conduct experiments and make observations, scientists are guided by theories from which hypotheses are logically deduced. Theories are not true by definition, however, Deductions from theory predispose broad inductions to validate their major premises. All knowledge of the world can only be achieved with some degree of confidence when we test our concepts in the world outside our own minds. Empirical science cannot produce the certainty demanded by those who identify all true knowledge with mathematics, but induction is the bedrock of science.

The third method of reasoning is abduction. Abduction starts with all available relevant evidence and proceeds to the most reasonable explanation for them but leaves open other possibilities. Peter Lipton offers an example of abductive reasoning in the form of Sherlock Holmes zeroing in on his arch-enemy, Professor Moriarty. Sherlock infers that Moriarty is guilty because that hypothesis best explains all the evidence gathered, such as fingerprints, bloodstains, and other such evidence. Lipton says that Sherlock's belief is not arrived at deductively: "The evidence will not entail that Moriarty is to blame, since it always remains possible that someone else was the perpetrator. Nevertheless, Holmes is right to make his inference, since Moriarty's guilt would provide a better explanation of the evidence than would anyone else's" (Lipton, 2000, p. 185).

The sum of the evidence that Lipton presents suggests that Moriarty is guilty "beyond a reasonable doubt," but not beyond all doubt. Unlike deductive reasoning whereby the conclusion is guaranteed by the axiom, the jury will have to reach the simplest and most logical conclusion it could draw from multiple lines of evidence. Science sets up such a decision-making process the same way that the Anglo-American legal system tests the guilt or innocence of the accused—it assumes that the person in the dock is innocent. This is called the null hypothesis. The assumption of innocence and null hypotheses are precautionary measures that require stringent evidence to reject it. Some of the evidence may carry little weight, but piling evidence upon evidence, upon evidence, will eventually build such a weighty case that few reasonable people will reject on purely rational grounds. When we have collected all relevant evidence, we test it against multiple competing hypotheses and make an inference to the best explanation.

A scientific theory must fit the known facts about a domain of inquiry into a logical pattern. It must also allow us to make predictions deduced from it about as yet unknown areas of that domain. The Anthropic Principle satisfies only the first of these criteria. It looks backward to coherently explain what is already known in terms of a purposeful universe created for us and can accommodate future discoveries. It does not ordinarily allow us to make predictions, although it has been claimed that some successful scientific predictions have been based on it. Anthropic reasoning is abductive; that is, a *post hoc* explanation of the facts already at hand. To give a simple example, you may observe that the street is wet and conclude that it has rained. It could also be wet if the street cleaners had just gone past or a water pipe had burst nearby. All three hypotheses (rain, street cleaners, a broken water pipe) have explanatory power; if any were true, it would explain why the street is wet. Intuitively, the rain hypothesis is much better than the others, especially if we seek further evidence. If we find that the roof is wet and that there is fresh water in the rain gutters, we can reject the other possibilities and conclude that the street is wet because it rained. We cannot, however, make a prediction based on our observation that it will rain on the same day next month.

Just as we can offer three different explanations of why we observed the wet street, we can offer three different explanations for anthropic fine-tuning. First, the universe is deliberately designed to allow for the emergence of intelligent life. Second, the universe is a big fluke and we are incredibly lucky that it has the right combination of physical, chemical, and biological conditions to lead to human life. Lastly, we can say that our universe is only one among trillions of other universes with physical laws that differ drastically from ours (the multiverse), and we just happen to live in one that is life-permitting. After surveying all available evidence, we have to choose from the three options by reasoning to the most sensible. These choices are not mutually exclusive or exhaustive; I simply present them in the way they are typically presented to us. As we shall see, there are many scientists who see the multiverse as a reality and indicative of God's endless creative power, and there are eminent theologians who accept (as do I) theistic evolution in which chance is God's way of imbuing nature with the power to create life. This was Darwin's position. My choice is simply one consistent with astronomer Robert Jastrow's poetic account of how science caught up with theology on the matter of creation in the Big Bang: "For the scientist who has lived by his faith in the power of reason, the story ends like a bad dream. He has scaled the mountain of ignorance; he is about to conquer the highest peak; as he pulls himself over the final rock, he is greeted by a band of theologians who have been sitting there for centuries" (Jastrow, 1992, p. 107).

The mind-melting improbabilities of the values of the myriad parameters necessary for a life-permitting universe we will be looking at utterly destroy the Copernican Principle. Something purposive is afoot in a universe in which intelligent beings can plumb its depths. The universe is very special, and so are its inhabitants. If we are special, does that mean that life is unique to this planet? This is a question that every philosopher of the heavens has asked at one time or another. The adjectives "special" and "unique" are similar but different. The former denotes value; the latter "one of a kind." I am asking not only are we and our planet special but also are we unique?

In the narrative of materialism, it is inexcusably egocentric to claim that Earth is the only planet in the cosmos with intelligent life. Of course, it really doesn't matter if we are not alone; God could have liberally salted the universe with intelligent life if that was His desire. His existence hardly rests on an answer to the question of Earth's uniqueness. At the time of writing, 4,843 confirmed exoplanets have been found, but none with all the requisites of a life-bearing planet. Astrophysicist Hugh Ross computes the probability of a planet falling within necessary parameters by chance at less than 1 in 10^{215} and states that: "fewer than a trillionth of a trillionth of a percent of all stars will have a planet capable of sustaining advanced life. Considering that the observable universe contains less than a trillion galaxies, each averaging a hundred billion stars, we can see that not even one planet would be expected, by natural processes alone, to possess the necessary conditions to sustain life" (Ross, 1994, pp. 169-170).

Many astrophysicists have reached a similar conclusion. Looking at the thousands of parameters that have to be just right for complex life, John Gribbin asks if it is likely to exist elsewhere in the universes and answers: "Almost certainly no, given the chain of circumstances that led to our existence" (2018, p. 99). Upon Stuart Taylor being awarded the prestigious Leonard Award for outstanding contributions to planetary science, he noted the staggering improbabilities of a myriad of things from the formation of the Milky Way to intelligent life on Earth and concluded: "When the remote chances of developing a habitable planet are added to the chances of both high intelligence and a technically advanced civilization, the odds of finding 'little green men' elsewhere in the universe decline to zero" (Taylor, 1998, p. 327). Then we have the opinion of astrobiologists Plaxco and Gross: "The range of values in Drake's parameters [an equation for estimating the probability of intelligent life beyond Earth] could adopt is so great, that despite the huge numbers of stars in the Universe, current scientific knowledge is entirely consistent with N=1. That is, Fermi [Enrico Fermi, the Italian-American Nobel laureate physicist] was right, and we are alone" (Plaxco and Gross, 2006, p. 247). There are an almost infinite number of ways in which the universe's evolution toward life could have gone wrong given the universality of the Second Law of Thermodynamics. This law presents far too many thermodynamic hurdles against the emergence of life because it tells us unequivocally that disorder (entropy) in the universe is always increasing. That we have a biocentric home on Earth constitutes remarkable evidence for the Strong Anthropic Principle.

CHAPTER TWO

God or Science?
The Conflict that Never Was

Science is the pinnacle of the human intellect. It is the tentative and self-correcting way by which we come to understand the natural world because it yields results verifiable. Science is a process by which answers lead to more questions; it feeds on ignorance, for what is already known is boring. But science does not claim to answer the big questions of existence. Christian theism has always done that without science, but in an age when science's evidential force is too strong to ignore, Christians must know how science points to God. To defend science is rather like defending mother, flag, and apple pie, for science has lifted humanity to such a level of health, freedom, and comfort undreamt of in the pre-scientific world. It is difficult to see how anyone can think ill of it, but some do, believing that it denies God. But affirming science does not imply denying God, as countless first-rate scientists attest. But some, such as Richard Lewontin, refuse to let God into their world and place all their faith in materialism. As he notes:

> It is not that the methods and institutions of science somehow compel us to accept a material explanation of the phenomenal world, but, on the contrary, that we are forced by our *a priori* adherence to material causes to create an apparatus of investigation and a set of concepts that produce material explanations, no matter how counter-intuitive, no matter how mystifying to the uninitiated. Moreover, that materialism is absolute, for we cannot allow a Divine Foot in the door. (Lewontin, 1977, np)

Scientists who won't allow a Divine Foot in the door must struggle to explain the most fundamental scientific questions, such as the origin of the universe and of life. Lewontin takes for granted that science is in a struggle with the supernatural and must stick to its materialistic guns. Atheists say they believe only in science and reason, and that Christians reject reason in favor of faith. But note that Lewontin insists that scientists *must* accept materialism—based on what? Not on the demands of the scientific method, but on faith! While some scientists and theologians believe there is a conflict between their respective domains, historian of science, Colin Russell, corrects them: "The common belief that the actual relations between religion and science over the last few centuries have been marked by deep and enduring hostility...is not only historically inaccurate, but actually a caricature so grotesque that what needs to be explained is how it could possibly have achieved any degree of respectability" (in Lennox, 2009, p. 28).

Russell is correct, so where did the notion of conflict begin? It may have begun with Pierre Simon Laplace and Napoleon Bonaparte. When Laplace present Napoleon with his 1798 *Treatise on Celestial Mechanics*, Napoleon asked him why, in a book on the mechanics of the universe, he had not mentioned the Creator, to which Laplace replied, "Sir, I had no need of that hypothesis" (in Keyser, 1915, p. 28). Laplace's remark was not a denial of God (he was a practicing Catholic) but was a statement affirming that his work was about *how* the universe works, not about *why* it exists or how it came to be. Astronomer Joseph-Louis Lagrange responded to Laplace, saying: "Ah, it [the God hypothesis] is a fine hypothesis; it explains many things." In turn, Laplace responded to Lagrange: "This hypothesis, Sir, explains in fact everything, but does not permit me to predict anything. As a scholar, I must provide you with works permitting predictions" (in Jennings, 2015, p. 59).

Laplace was correct; ultimately God explains everything, but it is true that the God hypothesis does not allow testable scientific predictions. But what if Napoleon had asked Laplace why we are here on this Earth, what is the purpose of life, or the biggest philosophical questions of all: "Why is there something rather than nothing?" These are questions science cannot answer, so God is hardly irrelevant if we wish answers to them. Science has discovered many things since LaGrange lived, and the more it discovers the more it finds mystery, and the more the mind is forced to contemplate God. Even if in the distant future we find the final equation of everything explaining *how* the universe works on both the macrocosmic and micro quantum scales, we would still not have answered *why*; why abstract mathematics has such an uncanny relationship with the physical universe, or we have an immaterial mind that brought the equation to light.

God's existence is not contingent on science's current inability to explain something or other in the natural world. Inserting God into a gap in scientific knowledge does theism a disservice, because if and/or when science does explain the phenomenon in question, "God-of-the-gaps" arguments play into the hands of atheists claiming that religion and science are in conflict, and makes God's work contingent on science's current limitations. Scientists readily acknowledge that the big questions of meaning are outside of their purview, so rather than inserting God into scientific gaps to argue from what we *don't* know about the workings of the universe, theistic scientists argue from what we *do* know, and how it logically leads to the transcendent Creator of the universe, who is the ground of all explanation.

It is a classical category mistake of confusing impersonal principles with personal agency to interpret Laplace's answer as implying that God is an unnecessary explanation. Science seeks answers to *how* God created the universe; theology searches for *why* He did. Mathematician and philosopher John Lennox uses the example of a Ford automobile to make this point. He says that an engineer, using the principles of internal combustion, can explain *how* the car works, but if he wanted to know *why* a car exists, he would have to invoke Henry Ford's agency in choosing to manufacture automobiles. Ford's agency is immaterial to explaining how a car works but necessary to explain why it exists. We need both explanations to have a necessary and sufficient explanation of the car. Likewise, to have a necessary and sufficient explanation of the universe, we need both science and the Creator. Lennox's evaluation of Laplace's remark is also revealing: "Considered as a serious observation, his remark could scarcely have been more misleading. Laplace and his colleagues had not learned to do without theology; they had merely learned to mind their own business" (Lennox, 2009, p. 46). "Their own business" is the business of science, which is pursued by even the most devout scientists without invoking God.

While some people see science and religion in conflict, others see the two domains as mutually supportive and still, others see them as totally separate domains, and thus cannot be in conflict. Conflict usually occurs when two systems of thought are contesting the same territory. Because this is so, paleontologist Stephen J. Gould, coined the term "non-overlapping magisteria" (NOMA) to refer to the notion that science and religion have legitimate authority in their non-overlapping domains of inquiry. Since these two magisteria do not overlap, there is no conflict between science and religion as long as both mind their own business; one addressing how questions, the other, why questions. As Gould put it, "we [scientists] study how the heavens go, and they [theologians] determine how to go to heaven" (1998, p. 31).

Non-overlapping magisteria are incommensurate domains of knowledge. "Incommensurate" and "contradictory" are not synonymous terms. Contradictory worldviews, such as the geocentric and heliocentric models of the solar system, can be reconciled with observations because both models speak the same language and seek the same truth. Incommensurable worldviews are radically incompatible in terms of such things as meaning, truth, or justification because the concepts of one cannot be clearly translated into the concepts of the other. In such a case, the two worldviews cannot logically be compared to expose contradictions, since there is no shared discourse. While modern science and modern religion may have different languages, they are not incommensurate. Nor are they in conflict unless one is determined to make them so.

Scientists of some repute have opined that Christianity and science are intimately connected. Einstein's familiar statement that "Science without religion is lame, religion without science is blind" is a case in point (in Huchingson, 2005, p. 149). Lord Willian Kelvin, who devised the absolute Kelvin temperature scale and formulated the second law of thermodynamics, wrote: "Do not be afraid of being free thinkers! If you think strongly enough you will be forced by science to the belief in God, which is the foundation of all religion. You will find science not antagonistic but helpful to religion" (in Smith, 1981, pp. 307-308). Another great man of science, astrophysicist Paul Davies: "It may seem bizarre, but in my opinion, science offers a surer path to God than religion ... science has actually advanced to the point where what were formerly religious questions can be seriously tackled" (1983, p. ix). Physicist Robert Russell believes that a sturdy bridge between Christianity and science exists: "In many ways, the bridge is now complete, and we can concentrate fully on the rich opportunities and challenges brought on by the flow of knowledge and vision in both directions across the bridge: the creative mutual interaction between theology and science" (2008b, p. 2). Physicist Own Gingerich also sees overlap, particularly as it applies to the issue of life's origin: "But it is a fallacy that there is no overlap of magisteria. In particular, the fascinating question of how and when we became human inevitably offers an overlap and potentially competing world views" (2014, p. 98).

The Christian Origins of Science

Certain *scientists* may be at war with God, but this is not the same as saying that *science* is at war with God. Atheists may be surprised to learn that so many of the advances in early science were made by men of God. Friar Roger Bacon is the father of the scientific method; Jesuit priest Roger Boscovich produced the precursor of atomic theory; the monk Gregor Mendel founded the science of genetics; Nicolas Steno, the father of geology was a priest; as was Jean-Baptiste Carnoy, the father of cell biology, and Georges Lemaître, the father of the Big Bang theory. As Albert Einstein asserted, the deep study of science offers a clearer path to God for the skeptic than arguments from theology: "The more I study science the more I believe in God" (in Holt, 1997, np). He also said: "Everyone who is seriously involved in the pursuit of science becomes convinced that a spirit is manifest in the laws of the Universe–a spirit vastly superior to that of man, and one in the face of which we with our modest powers must feel humble" (in Marsh, 2012, p. 72).

The spirit of science grew out of the Christian belief in a rational and orderly God who created us in His image. Einstein notes: "The eternal mystery of the world is its Comprehensibility. ... The fact that it is comprehensible is a miracle" (in Galison, Holton, and Schweber, 2008, p. 36). The fact that it is comprehensible and elegantly described by mathematics is proof that God wants us to understand His creation by using the intelligence He has gifted us. Nobel laureate chemist Melvin Calvin's understanding of the origin of the scientific conviction that the universe is orderly and knowable: "I seem to find it in a basic notion discovered 2,000 or 3,000 years ago, and enunciated first in the Western world by the ancient Hebrews: namely that the universe is governed by a single God and is not the product of the whims of many gods, each governing his own province according to his own laws. This monotheistic view seems to be the historical foundation for modern science" (in Lennox, 2009, p 20). This view is articulated in Solomon's prayer in Wisdom 7:17-22 in which he praises God for giving him the ability to do what he could to discover what God has hidden for us to find:

> For He gave me sound knowledge of what exists, that I might know the structure of the universe and the force of its elements, The beginning and the end and the midpoint of times, the changes in the sun's course and the variations of the seasons, Cycles of years, positions of stars, natures of living things, tempers of beasts, Powers of the winds and thoughts of human beings, uses of plants and virtues of roots—Whatever is hidden or plain I learned, for Wisdom, the artisan of all, taught me.

From the earliest days of Christianity, it was taught that reason is a unique gift of God, and that we must use this gift to come to know Him. The second-century theologian, Quintus Tertullian of Carthage informed us that: "Reason is a thing of God, inasmuch as there is nothing which God the Maker of all has not provided, disposed, ordained by reason—nothing which He has not willed should be handled and understood by reason" (in Coyne and Heller, 2008, p. 42). The Catholic Church founded the first true university—the University of Bologna—in 1088 and made mathematics and natural philosophy, or science as it's known today, compulsory parts of the education of anyone wanting to study theology. Christian theology thus created the fertile intellectual soil for science to grow. This hardly sounds like theologians with fear or distrust of science.

The great contribution of Christian theology to science lies in its conviction that there are laws of nature front-loaded by God at the beginning awaiting discovery. The thirteenth-century patron saint of science, Albertus Magnus, wrote: "It is the task of natural science not simply to accept what we are told but to inquire into the causes of things" (Kennedy, 1907, p. 265). Magnus also stated: "In studying nature we have not to inquire how God the Creator may, as He freely wills, use His creatures to work miracles and thereby show forth His power; we have rather to inquire what Nature with its immanent causes can naturally bring to pass" (Kennedy, 1907, p. 265). This does not have the sound of unreasoned faith. Sir Isaac Newton got the idea that there were absolute, universal, laws from the biblical doctrine that God created and ordered the universe in a rational way. Former atheist, geneticist, and physician, Francis Collins, finds God in his science: "I have found there is a wonderful harmony in the complementary truths of science and faith. The God of the Bible is also the God of the genome. God can be found in the cathedral or in the laboratory. By investigating God's majestic and awesome creation, science can actually be a means of worship" (Collins, 2007, np).

The Judeo-Christian exhortation to explore the fingerprints of God in the natural world is absent in the theology of other religions. Islam accepts God as the all-powerful Creator of the universe but does not believe that we can know Him through His creation. Islamic theology thus does not provide the fundamental assumptions necessary for the emergence of empirical science devoted to discovering natural laws. According to Rodney Stark: "Allah is not presented as a lawful creator but has been conceived of as an extremely active God who intrudes on the world as he deems it appropriate. Consequently, there soon arose a major theological bloc within Islam that condemned all efforts to formulate natural laws as blasphemy insofar as they denied Allah's freedom to act" (2003, p. 154). Thus, despite all the excellent scholarship of medieval Muslims in philosophy, mathematics, and medicine, the Islamic world never produced anything like Western science. If Christianity has dogmatically opposed science across the centuries, how is it that science only developed in Christian countries?

Materialism and Naturalism

There is no conflict between science and Christian theism, but there is a conflict between theism and materialism/naturalism. Atheistic scientists are committed to the metaphysics of ultimate meaninglessness; viewing everything that exists as the result of the combination of chance and the laws of physics. However, scientists can be thoroughgoing materialists in their daily work, but still reject the notion that the matter of nature they work with is all that there is. These scientists might be partial to Nobel laureate physicist Max Planck's words: "Both Religion and science require a belief in God. For believers, God is in the beginning, and for physicists, He is at the end of all considerations.... To the former, He is the foundation, to the latter, the crown of the edifice of every generalized world view" (1949, p. 184).

There are two forms of materialism: *methodological* and *ontological.* Methodological materialism does not carry any ideological baggage; it is simply a regulative principle of science. The most devout of scientists subscribe 100% to methodological materialism because it works, but they reject the notion that the material world exhausts all reality. As Allan Sandage, the "grand old-man of cosmology" noted, many never take the extra step: "Those that are content in every part of their being to live as materialistic reductionalists (as we all do as scientists in the laboratory, which is the place of the practice of our craft) will never admit to a mystery of the design they see, always putting off by one step at a time, awaiting a reductionalist explanation for the present unknown" (1985, p. 53). Thus, ontological materialism goes beyond a working assumption to claim that science has exclusive access to the truth and is the only way to access reality. It maintains that there is nothing beyond the materialist realm of being, and: "Whatever knowledge is attainable must be attained by scientific methods; and what science cannot discover, mankind cannot know" (Russell, 1935, p. 243). In this view, we should put up the shutters on all university departments other than the hard sciences, because only science is capable of providing comprehensive knowledge of all reality.

The problem with materialism/naturalism comes when we jump from a working assumption to a comprehensive philosophy; that is, the assumption that there is nothing beyond the realm of the natural. The inherent atheism of ontological materialism/naturalism is made transparent by the infamous Madalyn Murray O'Hair, who asserts that: "Atheism is based upon a materialist philosophy which holds that nothing exists but natural phenomena. There are no supernatural forces or entities, nor can there be any. Nature simply exists" (in Weitnauer, 2013, p. 28). A chain of rational reasoning about human existence that ends abruptly when atheists arrive at the beginning of the universe with "Well, that's just the way it is," wants us to accept the notion that the brute fact of existence is ultimately meaningless. This is not an explanation of anything; it is a discussion ender.

Unlike methodological materialism, ontological materialism can neither be affirmed nor denied by science and thus is just as metaphysical as theism. Theism affirms a reality beyond matter which also cannot be conclusively affirmed or denied. The claim that there is a conflict between science and theism should thus read that there is a struggle between *ontological* materialism and the supernatural. But ontological materialists seem to be in the minority among the very best scientists. Baruch Shalev's book documenting the religious views of all 719 Nobel Prize winners from 1901 to 2000 found that most were friendly to theism (Shalev, 2003). Only 10.5% of Nobel laureates fell into the atheist or agnostic category. Most (35.2%) of the Nobel winners falling into the category were winners in literature; only 4.7% of winners in physics did so. This may be the reason prize-winning mathematical physicist Robert Griffiths said: "If we need an atheist for a debate, we go to the philosophy department. The physics department isn't much use" (in Kainz, 2010, p. 21).

Materialism and naturalism are often considered synonymous, but there are differences. The most dogmatic of the two in its ontology is materialism. Materialist ontology avers that all existence is matter, only matter, the physical "stuff" you can see, touch, measure, and manipulate, is real, and that there is no metaphysical reality. Naturalism shares with materialism its denial of causal mechanisms outside the natural, which means that it denies the supernatural, but naturalists also deny that all effects must have material causes. The philosophical position that naturalists and materialists agree on is that there is nothing separate, above, or prior to the natural world. Some naturalists, such as Max Planck, the father of quantum mechanics affirm the existence of a non-material mind, and even that the universe is made of "mind stuff":

> As a man who has devoted his whole life to the most clear-headed science, to the study of matter, I can tell you as a result of my research about atoms this much: There is no matter as such. All matter originates and exists only by virtue of a force which brings the particle of an atom to vibration and holds this most minute solar system of the atom together. We must assume behind this force the existence of a conscious and intelligent Mind. This Mind is the matrix of all matter. (In Olsen, 2013, p. 382)

Planck's belief at the most fundamental level that the universe is made of "mind stuff" is shared by a number of other physicists. Astrophysicist Sir James Jeans famously wrote: "The stream of knowledge is heading towards a non-mechanical reality; the Universe begins to look more like a great thought than like a great machine. Mind no longer appears to be an accidental intruder into the realm of matter... we ought rather hail it as the creator and governor of the realm of matter" (Jeans, 1930, p. 137). Another great astrophysicist, Sir Arthur Eddington, wrote: "The universe is of the nature of a thought or sensation in a universal Mind" (in Schafer, 2006, p. 509). When we think about the almost infinite divisibility of matter, we have to wonder what it ultimately is; does it point to John 1:1 "In the beginning was the Word, and the Word was with God, and the Word was God"?

For the ontological materialist, however, mental phenomena are illusionary and merely reflect electrical energy moving stuff around in the brain. For instance, Francis Crick has infamously written that: "'You,' your joys and your sorrows, your memories and your ambitions, your sense of personal identity and free will, are in fact no more than the behavior of a vast assembly of nerve cells and their associated molecules. Who you are is nothing but a pack of neurons" (Crick, 1994, p. 3). Crick might have traveled further along the materialist road to say that you are nothing but a pack of quarks because neurons are ultimately made of them. Mental phenomena require a physical platform called the brain, but the mind cannot be reduced to it without the remainder. Crick's words come from his book, *The Astonishing Hypothesis*. It is indeed astonishing, and one wonders why he didn't subtitle it *The Zombie Within*.

What would Crick make of that delicious feeling we call romantic love? He might say that neuroscientists look at the chemical and physical activity lighting up the brain's pleasure centers when people are in love, and that is what we call love. But can we reduce the intoxication of romantic love to the soup and sparks of brain activity? These material things do not come remotely close to explaining why Romeo fell in love with Juliet; they merely tell us what happened in his brain when he did. Love is an example of "top-down" causation because love comes first, and only then came the brain's soup and sparks. The soup and sparks represent all the emotions that Juliet conjured up in Romeo's brain and are necessary for love to exist, but the physical product is a consequence of the mental process and not the other way around.

Chapter Three

The Miracle of Mathematics

Mathematics and Beauty

Galileo Galilei once wrote: "Mathematics is the language in which God has written the Universe" (in Ilić, Stefanović, and Sadiković, 2018, p.124). Early scientists such as Copernicus, Galileo, Kepler, and Newton knew that the universe was capable of mathematical description because a rational God fashioned it in a rational way. Mathematics is the sturdy backbone of all science and the unreasonable effectiveness of abstract mathematics in describing empirical reality has given it a mystical aura. As Nobel laureate physicist Roger Penrose said of it: "There is something absolute and 'God given' about mathematical truth" (2016, p. 146). And another British Nobel laureate, physicist Paul Dirac, opined: "God is a mathematician of a very high order and He used advanced mathematics in constructing the universe" (in Varghese, 2013, p. xviii). Many Christian theologians and mathematicians from Augustine (354 – 430 AD) onwards, have believed that mathematics is in the mind of God, and when we discover these eternal mathematical truths we have insight into the mind of God, or at least, we see signs of the Divine.

Most mathematicians and many physicists are unabashed Platonists about mathematics. That is, they view mathematics as having an ontological reality that is discovered rather than invented. Mathematical truths represent the real world in abstract symbols and have been amazingly successful in doing so. Mathematical equations may represent some aspect of reality already known to the senses but not well understood at the time, such as motion and gravity described by Newton, or electricity and magnetism described by Maxwell. Newton developed calculus as a practical matter because he needed a tool to understand movement and force, but mathematicians love to play around with shapes and equations with no immediate practical use in mind. For instance, German mathematician Bernhard Riemann's work in non-Euclidean geometry in the 1850s found no practical application at the time since he was not trying to describe some physical phenomena. However, his mathematical foundation for the four-dimensional geometry of space-time turned out to be exactly the tool Einstein needed to formulate his famous 1915 theory of relativity more than 50 years later.

Of course, Newton was aware of force and momentum and Maxwell knew of electricity and magnetism before they shined their mathematical torches on them, but mathematics may predict something not known at the time that is amenable to experimental testing. An example of this is the prediction Einstein made from his general theory of relativity that light is bent by gravity. This was thought to be impossible because photons are massless and massless things can't bend, but Einstein's equations revealed that gravity bends spacetime. Sir Arthur Eddington empirically proved Einstein right in 1919 by observing and photographing the bending of light in a solar eclipse. More recently, the Higgs boson was uncannily predicted by mathematics almost 50 years before it was discovered in 2012. Numerous other examples justify mathematical models as faithful representations of empirical reality.

In many ways, mathematics is a mystery whose depth and beauty are bottomless. Many people have expressed the notion that mathematical equations are things of beauty, but when I have tried to convey this notion to my statistics students, I get a sea of rolling eyes because they see only the mystery. Students everywhere, it seems, call statistics "sadistics" and ask themselves: "How can anyone think that this torture is beautiful?" Nevertheless, math is a thing of beauty, particularly if it is, as Galileo said, the language by which God wrote the Universe. As Bertrand Russell so splendidly put his opinion of the beauty of mathematics:

> Mathematics, rightly viewed, possesses not only truth, but supreme beauty, a beauty cold and austere, like that of sculpture, without appeal to any part of our weaker nature, without the gorgeous trappings of painting or music, yet sublimely pure, and capable of a stern perfection such as only the greatest art can show. The true spirit of delight, the exaltation, the sense of being more than Man, which is the touchstone of the highest excellence, is to be found in mathematics as surely as in poetry. (1919, p. 60)

An example of mathematical beauty is the notion of imaginary time, developed by the prodigious intellect of the late Steven Hawking. To get the gist of what Hawking is getting at, imagine a straight horizontal line describing the arrow of time from its beginning in the Big Bang into the distant future. This is real time. Hawking's imaginary time, which he says is as "real" as real time, is imagined as a vertical line intersecting real time at right angles. From where to where this timeline is traveling? Hawking admits he has no idea. This imaginary time is said to have existed before the Big Bang and was always there and existed in a "bent" state. Hawking knows that the laws of physics cannot explain anything before the Big Bang because there was no "before" and that his imaginary time has no meaning in reality, but he has conjured it up by virtue of his almost unmatched grasp of mathematics.

To arrive at his model of imaginary time, Hawking uses imaginary numbers such as the square root of -1. Imaginary numbers do not have a tangible value; you can't use them to figure your grocery bill but they are "real" in the sense that they are used in higher mathematics and physics. Just as imaginary time is perpendicular to real time, imaginary numbers are perpendicular to the familiar horizontal number line. Hawking thus uses imaginary numbers to invent imaginary time which is something much admired by mathematicians for its beauty. He makes it clear that he is unconcerned about whether or not extra dimensions such as "bent" time are real as long as they appear in the math: "I have been reluctant to believe in extra dimensions. But as I am a positivist, the question 'Do extra dimensions really exist?', has no meaning" (Hawking, 2001, p. 54). So, Hawkins asserts that to ask if something he says exists does, in fact, exist, is philosophically meaningless. This is a strange idea, to say the least. Imaginary time has no more meaning in everyday empirical reality than the imaginary bank account I dream of, but it cannot be denied that mathematically his model is beautiful.

But mathematical beauty does not confer reality on the universe; God does that. In his book, *A brief history of time*, Hawking asks: "What is it that breathes fire into the equations and makes a universe for them to describe? The usual approach of science of constructing a mathematical model cannot answer the questions of why there should be a universe for the model to describe. Why does the universe go to all the bother of existing?" (1988, p. 174). This is the philosopher's "why" question, not the scientist's "how" question, but it shows that while equations describe aspects of empirical reality, they do not confer reality on it any more than an architect's plans for a building are the building. A few pages later, Hawking notes that the question he posed earlier is of tremendous importance: "If we find the answer to that, it would be the ultimate triumph of human reason—for then we should know the mind of God" (Hawking, 1988, p.193). Hawking was an atheist so he meant this only metaphorically, but it was God who "breathes fire into the equations and makes a universe for them to describe."

Beauty does not always equal truth because sometimes equally beautiful equations may clash violently, particularly if applied to seemingly incommensurate domains, such as general relativity and quantum mechanics. Mathematical physicist Don Page notes this: "different mathematical structures can be contradictory, and contradictory ones cannot co-exist. For example, one structure could assert that spacetime exists somewhere and another that it does not exist at all...these two structures cannot both describe reality" (Page, 2007, p. 424). There are things in nature that cannot always be expressed in stringent formulae because of the uncertainties (chaos) involved. Herbert Dingle, a formidable mathematician, tells us how almost anything imaginable can be done with mathematics: "In the language of mathematics we can tell lies as well as truths, and within the scope of mathematics itself there is no possible way of telling one from the other. We can distinguish them only by experience or by reasoning outside the mathematics, applied to the possible relation between the mathematical solution and its physical correlate" (Dingle, 1972, pp. 31-32). Furthermore, Albert Einstein notes: "As far as the laws of mathematics refer to reality, they are not certain; and as far as they are certain, they do not refer to reality" (1923, p. 28). Mathematics must conform to reality, not reality to mathematics. Nevertheless, mathematics remains "The queen of the sciences," and only displays unreality in the hands of those who believe that mathematical beauty equals truth in all instances.

Probability and its Limit

A great deal of scientific evaluation involves calculating probabilities. We cannot always say that given X, Y *will* occur; we say that given X, Y has a certain *probability* of occurring. As noted earlier, when scientists make scientific observations, they are like jury members in a criminal trial instructed not to convict the offender unless convinced that he is guilty beyond a reasonable doubt. The judicial system assumes innocence until the facts indicate otherwise. Likewise, scientists have a presumption of "innocence"—that X does *not* cause Y; or the null hypothesis. Scientists will not reject an assumption that X does not cause Y unless probability calculations tell them that they can do so "beyond a reasonable doubt." A very liberal probability would be 0.01, meaning that the research hypothesis that observation X *does* cause Y could be wrong by chance 1 out of every 100 times the experiment is conducted and the outcome measured. If a computed probability is greater than 0.01, the null hypothesis is retained. If 0.01 seems too generous, a more conservative one is 0.001, or one chance in a thousand. Rare things obviously happen; we just have to determine how likely or unlikely they are to happen using statistics.

Many of the probabilities we will encounter in this book are so astronomically small, that we are justified in thinking them impossible. What is the point at which something improbable becomes impossible; not just beyond a reasonable doubt, but beyond all possible doubt? Mathematician William Dembski has computed the absolute limit of probability using three estimates from astrophysics: (A) the estimated number of atoms in the known universe (10^{80}); (B) Planck time (10^{-45}), and (C) the number of seconds since the Big Bang at the time of his calculations (10^{25}). Planck time sets an absolute limit on the rate at which elementary particles can transition from one state to another. From these absolute limits—all existing matter; the shortest period of time, and the total time available in which everything in the universe has happened, Dembski concludes: "If we now assume that any specification of an event within the known physical universe requires at least one elementary particle to specify it and that such specifications cannot be generated any faster than the Planck time, then these cosmological constraints imply that the total number of specified events throughout cosmic history cannot exceed $10^{80} + 10^{-45} + 10^{25} = 10^{150}$" (2004, pp. 84-84). This is by far the most conservative estimate of the probability boundary. It completely exhausts all probability resources since it includes the sum of all the atoms in the universe, all the seconds since the universe began, and the fastest possible time in which an event can occur.

Exponential numbers such as 10^{150} are a lot larger than they seem to those not too familiar with exponential language. Dembski's probability boundary is 1 followed by 150 zeros. To put it in perspective, one million is 10^6, a billion is 10^9, and a trillion is 10^{12}, or 1 followed by just 12 zeros, and remember scientists are willing to use 10^3 (1 in 1,000) as a conservative level to rule out chance (while not denying that chance remains a possible explanation for an observation). Numbers larger than a trillion are difficult for us to wrap our minds around, but astrophysicist Hugh Ross provides us with a visual image that helps us to understand the immensity of the number 10^{37}, which, while huge, is still vastly smaller than Dembski's 10^{150}. Written out, 10^{37} is: 10,000,000,000,000,000,000,000, 000,000,000,000,000, or ten undecillion. Ross' visual image is very useful for grasping such a number.

> Cover the entire North American continent in dimes all the way up to the moon, a height of about 239,000 miles (in comparison, the money to pay for the U.S. federal government debt would cover one square mile less than two feet deep with dimes). Next, pile dimes from here to the moon on a billion other continents the same size as North America. Paint one dime red and mix it into the billions of piles of dimes. Blindfold a friend and ask him to pick out one dime. The odds that he will pick the red dime are one in 10^{37}. (Ross, 1993, p. 115)

The Golden Ratio

Chemists Boeyens and Comba note that: "One of the most mysterious observations in Nature is the appearance of a single parameter that determines the macroscopic structure of a large variety of apparently unrelated objects... This ubiquitous parameter, known as the golden ratio, has also been called the divine proportion"

(2013, p. 1). The Golden Ratio is a number designated by the Greek letter φ (phi) and is thought to reflect the ideal proportions of nature. The mind easily finds beauty in this ratio, because it is painted in the entire cosmos from macroscopic spiral galaxies to microscopic quantum phenomena. Mario Livio tells us that great mathematical minds from Pythagoras in ancient Greece to Nobel laureate Roger Penrose in modern Britain: "have spent endless hours over this simple ratio and its properties. ... Biologists, artists, musicians, historians, architects, psychologists, and even mystics have pondered and debated the basis of its ubiquity and appeal. In fact, it is probably fair to say that the Golden Ratio has inspired thinkers of all disciplines like no other number in the history of mathematics" (Livio, 2003, p. 6). Medical researchers Ilić, Stefanović, and Sadiković, also reflect on this fascinating number:

> The intellectual elite of wide domains of scientific interests has been deeply fascinated by the Golden ratio for more than 24 centuries. In fact, it is practically impossible to specify all of the organic and inorganic living structures where this extraordinary number is hidden. Furthermore, according to the scientific point of view, new aspects of the living nature in which the pattern of the gorgeous mathematics have appeared and are still being revealed. There are specific, precise, and accurate relations among the existing forms in organic and inorganic aspects of nature which seem to be in perfect proportion and therefore they are called "Divine." In order to be attributed as "Divine", the proportion needs to have specific symmetry, beauty, and harmony. (2018, p. 125)

Although imbued with an almost mystic aura, mathematically it is simple to derive. The Golden Ratio (φ) is found when we divide a line into two parts, a and b, such that the ratio of the sum of the numbers (a + b) divided by the larger number (a) is equal to the larger number divided by the smaller number (a/b): $[(a + b)/a = a/b = \varphi]$. In other words, the whole length (a + b) divided by the long part (a) is equal to the long part (a) divided by the short part (b). Only two values will give you this ratio: 61.8 and 38.2. Thus: (a + b)/a = 100/61.8 =1.618.... and a/b = 61.8/38.2 = 1.618.... The dots after the number 8 indicates that the actual value of φ goes on forever, just like the irrational number π. A rectangle (of whatever size) formed by the two numbers of the ratio provides the most esthetically pleasing proportion for a painting, or anything else rectangular. In fact, almost all of the great paintings gracing the world's museums, and almost all great buildings show a close approximation to this "divine proportion" (Thapa and Thapa, 2018). We don't know, of course, whether the painters or architects were aware of the Golden Ratio (many were), or if it was just a matter of their innate sense of esthetic beauty that chose the proportion.

Figure 3.1. The Golden Spiral and Rectangle Formed by the Fibonacci Sequence

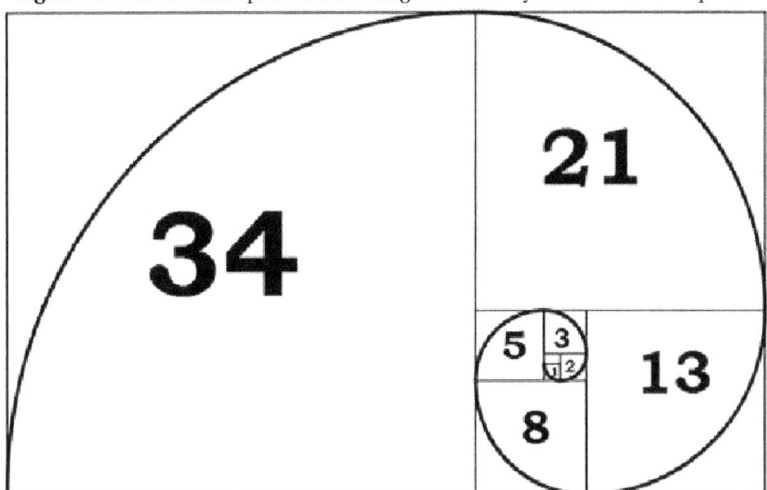

The Golden Ratio and the Fibonacci Sequence

The Fibonacci sequence is tightly connected to the Golden Ratio. The Fibonacci sequence was developed by the Italian mathematician Leonardo Fibonacci (b.1170-d.1250 AD). Ilić, Stefanović, and Sadiković write of many of God's imprints, and state: "One of those 'God's imprints' in the world that surrounds us is the Fibonacci Sequence, which represents more likely a starting point in revealing certain universal formula of life" (2018, p. 124). The Fibonacci sequence is a sequence of numbers in which each number is the sum of the previous two; viz: 0, 1, 1, 2, 3, 5, 8, 13, 21, 34, 55, 89, 114, 233, 377…and so on to infinity. The ratio of adjacent numbers in the sequence approaches the Golden Ratio after 8/5 = 1.6. The larger the numbers in the sequence, the closer the ratio is to the Golden Ratio. This seems deceptively simple, but mathematicians find many beautiful symmetries involving complex equations when they play around with the sequence. In fact, there are so many fascinating mathematical aspects to be found within them that there is even a scientific journal called *The Fibonacci Quarterly* in which mathematicians and other scientists can display their work. Mathematician S. Sinha calls the Fibonacci numbers "nature's numbering system" and gives a few examples of where they appear:

> The Fibonacci numbers are Nature's numbering system. They appear everywhere in Nature, from the leaf arrangement in plants, to the pattern of the florets of a flower, the bracts of a pinecone, or the scales of a pineapple. The Fibonacci numbers are therefore applicable to the growth of every living thing, including a single cell, a grain of wheat, a

hive of bees, and even all of mankind. Nature follows the Fibonacci numbers astonishingly. (2017, p. 13)

Using the Fibonacci sequence, we can build what is called the golden spiral. This spiral pattern is seen in countless instances in nature from galaxy formations to hurricane patterns, from a sunflower's spiral florets to leaf branching in trees, from snail shells to the DNA double helix, and from facial patterns to bodily proportions. Note how the spiral is built from squares in Figure 3.1. We begin with the innermost squares (the Fibonacci sequence 1, 2, 3, and 5), each of which forms a rectangle divided in two according to the Golden Ratio (the smaller one composed of 1, 2, and 3) and the larger one (5). The innermost square then becomes part of the rectangle composed of the squares marked 8 and 13 divided according to the Golden Ratio. We then draw successively larger squares with each two adjacent ones forming a rectangle. There's no limit to the squares that can be drawn, nor is it necessary to go as far as numbers 21 and 34. As well as the spiral being designated as "golden," the resulting rectangle is known as the golden rectangle. If you go to the Math is Fun website at https://www.mathsisfun.com/numbers/golden-ratio.html you can play around with different patterns of flower petals and see that only φ works; other irrational numbers such as π and *e* result in a big mess.

There are other wonders that issue from the Fibonacci sequence, such as the Fibonacci cascade that have implications for medicine and many other disciplines. For instance, Yalta, Ozturk, and Yetkin, writing in the *International Journal of Cardiology*, note that: "sequential multiplication by 1.618 or division by 0.618 (starting from 1) gives rise to a cascade of numbers (1, 1.618, 2.618, 4.236, 6.854, 11.09...) termed as 'Fibonacci Cascade' that was previously proven to be associated with branching patterns of human coronary arteries and a variety of botanic structures in nature" (Yalta, Ozturk, and Yetkin, 2016, p. 108). Another team wrote similarly: "One of the most notable concepts that might be drawn from these studies is that stimulation, contraction diameters and pumping pressure of heart obey the rule of Fibonacci series or golden proportions. And it seems that this harmony has been programmed and regulated by subtle forces or God. ... Explaining this relationship will help us to figure out the regularity of universe which is designed by God" (Ozturk, Yalta, and Yetkin, 2016, p. 145).

We began this chapter with top-tier scientists noting that God is a mathematician of the highest order. God is a God of order and truth, so anything orderly and truthful is a fact about God. But order and truth are not characteristics God displays, they are qualities that define his nature, and thus they flow from His nature. Mathematics displays the highest level of order and truth that humans possess. Stressing the complexity of the mathematics involved in quantum mechanics, Nobel laureate physicist Eugene Wigner writes: "It is difficult to avoid the impression that a miracle confronts us here quite comparable in its striking nature to the miracle that the human mind can string a thousand [mathematical] arguments together without getting itself into contradictions or the two miracles of the existence of the laws of nature and of the human mind's capacity to divine them" (1990, p. 7). Thus, there are many other forms of mathematics much more complex than the Golden Ratio and the Fibonacci sequence that lead scientists to contemplate the mystery of existence, but: "The Fibonacci sequence, concept, and theory represents a monumental complex that significantly exhibits a correlation and peculiarity far too precise and exact to occur by random processes, thereby advocating for the existence of an Intelligent Designer" (Dean-Lindsey, 2021, p. 3).

CHAPTER FOUR

Finding God in the Micro World: The Standard Model of Particle Physics

The Micro World

As science has poked around Mother Nature, she has divulged many secrets. Science has been so successful in this endeavor that perhaps one day it will be able to explain everything in nature. But don't confuse the explanatory power of science with God's creative power. They are entirely different categories; the first is the tool, the second is the toolmaker. Science discovers nature's laws, but it cannot explain why they exist. Nevertheless, atheists assume that Christians think of God as a placeholder filling a gap until a scientific explanation is found and argue that science has rendered God unnecessary. This is a God-of-the-gaps argument that no thinking Christian subscribes to. God does not live between science's gaps, His presence is felt in what we know, not in what we don't. As the gaps in scientific knowledge close, far from squeezing God out, science reveals wonders that could only be the handiwork of an infinite Mind. This is the opinion of Nobel laureate physicist Joseph J. Thomson: "As we conquer peak after peak we see in front of us regions full of interest and beauty, but we do not see our goal, we do not see the horizon; in the distance tower still higher peaks, which will yield to those who ascend them still wider prospects, and deepen the feeling, the truth of which is emphasized by every advance in science, that 'Great are the Works of the Lord'" (in Singh, 2004, pp. 361-362). We begin our exploration of God's great works at the microscopic level of atoms and their constituent parts.

Atoms are the building blocks of everything material. There are 92 kinds of natural atoms we call elements. An element is a pure substance that contains only one type of atom, but its bulk properties depend on how its atoms are connected. For instance, pure carbon is made from only one type of atom, but diamonds and graphite are vastly different physical forms of carbon. Elements increase in the periodic table according to the number of protons and neutrons in the nucleus. Isotopes are atoms that have the same numbers of protons but different numbers of neutrons. A proton is positively electrically charged balanced by an electron with an equal negative charge to make the atom electrically neutral. A neutron has no charge. Atoms that are not electrically neutral—they have gained or lost electrons—are called ions. The mass of an electron is about 1/2000 the mass of a proton or neutron and orbits far away from the nucleus. It has been estimated that if we could magnify the hydrogen nucleus (one proton, one electron, no neutron) to the size of a basketball, its electron would be about two miles away. This accounts for the notion that atoms are mostly empty space, which is not entirely true. The electron is smeared out into a cloud of possibilities called an orbital in superposition.

The "average" atom measures 100 picometres; that's many trillionths of an inch. Physicists have estimated that there is just about half the number of atoms in a single grain of sand, as there are grains of sand on all the Earth's beaches. Another way of looking at it is that if we blew an atom up to the size of a tennis ball and then blew the tennis ball up to the same extent, the ball would be about the size of the Earth. This gives us some idea of the insanely small size of an atom. Its constituent parts are even smaller, and they themselves, with the exception of electrons, which are fundamental particles, consist of even smaller particles.

Figure 4.1 presents the Standard Model of Particle Physics. The particles are elementary because they cannot be broken down any further. The building blocks of atoms are fermions, the matter particles that are "stuck" together by the bosons. Bosons are the carriers of the fundamental forces of nature. The two types of fermions are quarks and lepton. Protons and neutrons are made of three quarks; protons contain two up quarks and one down, and neutrons have two down quarks and one up. Unlike quarks that are subject to all of the fundamental forces, leptons are subject to all forces except the strong force. We don't have to get into the complications of mass, charge, and spin; suffice to say that these values are used by particle physicists to explore the behavior of protons and neutrons. We will discuss this particle menagerie further when we discuss the four fundamental forces of nature, but first a closer look at the Higgs boson.

The Higgs boson is a special kind of beast in this menagerie, being a scalar rather than a gauge boson. Gauge bosons are responsible for transmitting the forces that control how matter particles interact; the Higgs boson is different in that it interacts with elementary matter particles to give them mass. The Higgs boson was theorized first in 1964 when scientists, including Peter Higgs after whom it is named, were perturbed that their theories predicted a universe without mass. They believed that all particles in the early universe had no mass and traveled at the speed of light and that there must be a quantum field that gives particles their mass, which they later called the Higgs field. The Higgs field is a field of oscillating energy that pervades the universe, and when particles pass through it, they are slowed down and gain small quantities of that energy. If they interact with the field long enough, they accumulate energy as mass. There are only two known particles that zip through the Higgs field without interacting and are thus massless: photons that carry the electromagnetic force and gluons that carry the strong force. Quantum theory informs us that all fields have particles associated with them, so the Higgs field must have its own force-carrying particle produced through the quantum excitation of the Higgs field. The Higgs boson (whimsically nicknamed the "God Particle") is the quantum of energy (like a photon is a single quantum of electromagnetic energy) with which the Higgs field interacts with other particles and is produced through the quantum excitation of the Higgs field.

Figure 4.1. Standard Model of Elementary Particles

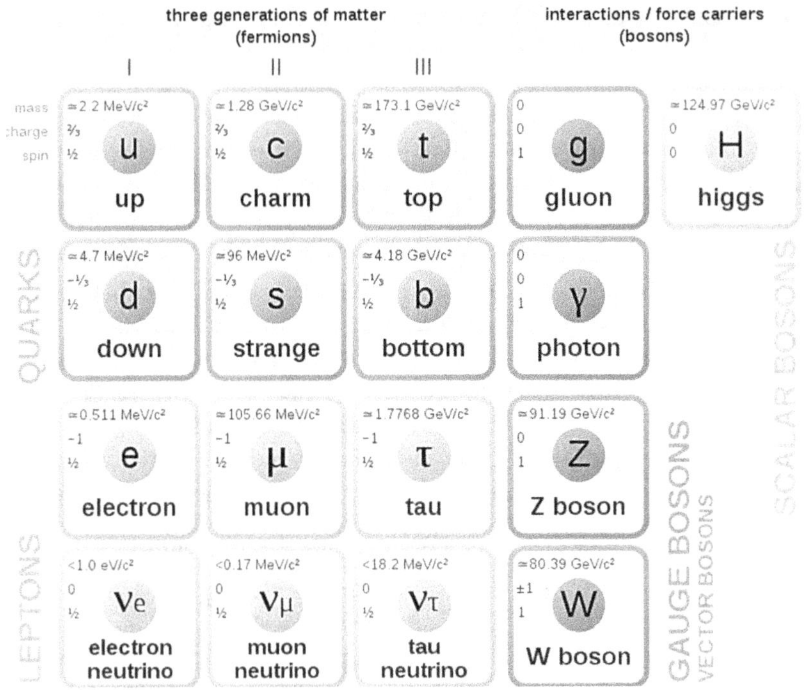
three generations of matter
(fermions)
interactions / force carriers
(bosons)
I
II
III
mass
charge
spin
≃2.2 MeV/c²
⅔
½
u
up
≃1.28 GeV/c²
⅔
½
c
charm
≃173.1 GeV/c²
⅔
½
t
top
0
0
1
g
gluon
≃124.97 GeV/c²
0
0
H
higgs
≃4.7 MeV/c²
-⅓
½
d
down
≃96 MeV/c²
-⅓
½
s
strange
≃4.18 GeV/c²
-⅓
½
b
bottom
0
0
1
γ
photon
≃0.511 MeV/c²
-1
½
e
electron
≃105.66 MeV/c²
-1
½
μ
muon
≃1.7768 GeV/c²
-1
½
τ
tau
≃91.19 GeV/c²
0
1
Z
Z boson
<1.0 eV/c²
0
½
νe
electron neutrino
<0.17 MeV/c²
0
½
νμ
muon neutrino
<18.2 MeV/c²
0
½
ντ
tau neutrino
≃80.39 GeV/c²
±1
1
W
W boson
QUARKS
LEPTONS
GAUGE BOSONS
VECTOR BOSONS
SCALAR BOSONS

The search for the Higgs had to await the building of the largest machine ever built: The Large Hadron Collider (LHC). The LHC is a 17-mile ring of superconducting magnets spanning the borders of Switzerland and France and run by CERN. It operates by sending protons along a number of accelerating structures in opposite directions at within decimal points of the speed of light, smashing them together, and statistically analyzing the resulting debris. The difficulty in finding the Higgs boson is that it decays in 10^{-22} seconds, which led William Lane Craig to comment that the reason it was labeled the "God particle" is because like God, it underlies every physical object that exists, and is very difficult to detect. Finding the boson is effectively the same as proving the existence of the Higgs field.

The discovery of the Higgs capped 50 years of spectacular success for the Standard Model of particle physics. Physicist Harry Cliff tells us that physicists regard the Standard Model as highly "unnatural" because of the large number of particles and forces that are balanced on a razor's edge such that changing any of the values: "you rapidly find yourself living in a universe without atoms. This spooky fine tuning worries many physicists, leaving the universe looking as though it has been set up in just the right way for life to exist" (Cliff, 2013, np). "Naturalness" is the prohibition of anthropic fine-tuning. Parameters that don't naturally emerge from a theory are called finely tuned, and physicists don't like that because they want their universe not to appear contrived. The Higgs boson has so much unnaturalness about it that particle physicist Euan McLean says that it has led some physicists to panic about its "spooky" nature. The problem is that its measured mass is fine-tuned by multiple trillions of degrees more than the Standard Model predicts: "It seems like, to generate a universe remotely like the one we live in, nature needs to decide on a parameter m_0 [the bare mass of a particle before interacting with a field to give it its measured mass m] highly tuned to 33 decimal places" (McLean, 2017, np).

Particle physicist Michael Strauss finds nothing spooky in this; rather, he sees the Higgs as providing insights into the existence of God. He marvels at the fact that physicists sat down in 1964 and came up with mathematical calculations that predicted that the Higgs field should exist, and then 48 years later it was found. The fact that we can describe the universe mathematically leads Strauss to opine that all explanations other than God are inadequate to explain our highly complex universe. He says of the Higgs particle: "Though it may not be properly 'The God Particle,' the mathematical description and complexity of our universe, along with its actual existence, gives a clear indication of a true deity who has designed and created what we now have the privilege to observe and study" (Strauss, 2017, np). Those of a contrary position consider it another filled gap pushing God out of the picture. In response to that, John Lennox wrote: "The meaning of the universe will be found where Newton and Clerk Maxwell found it: in God. So, what can we say about the Higgs boson? Simply this: God created it, Higgs predicted it, and CERN found it" (2012, np).

The Fundamental Forces of Nature

Everything in the universe is governed by just four forces. gravity, electromagnetism, and strong and weak nuclear forces. The four fundamental forces all work with each other to attain a universe fit for life. As Stephen Hawking and Leonard Mlodinow note, these forces are so fine-tuned that even the slightest variation in their values and the universe would not exist: "The emergence of the complex structures capable of supporting intelligent observers seems to be very fragile. The laws of nature form a system that is extremely fine-tuned, and very little in physical law can be altered without destroying the possibility of the development of life as we know it. Were it not for a series of startling coincidences in the precise details of physical law, it seems, humans and similar life-forms would never have come into being" (Hawking and Mlodinow, 2010, pp. 160-161). Hawking and Mlodinow call the laws nature and the phenomena they describe "startling coincidences." How many coincidences are needed for them to admit that something intentional is going on?

Gravity

Gravity is the force that operates throughout the universe. It is the force that gathered the material of the Big Bang and made it coalesce in stars and planets. The continued existence of stars is a balancing act between the force of gravity pushing in and the pressure from the explosive gases produced by burning hydrogen pushing out. If gravity was any stronger a star would collapse; any weaker and there would be no stars at all. Gravity is very powerful at the level of big things like stars, but it is by far the weakest of the four forces. Unlike the other fundamental forces, gravity does not have a known force carrying it. Although there is a hypothesized carrier called a graviton, because they would interact extraordinarily weakly with matter, we don't have equipment sensitive enough to detect one. All objects are subject to gravity, the strength of which depends on an object's mass, and how close it is to other objects. We are constantly being pulled down towards the Earth, but because we have so little mass, we don't feel it unless we fall. Without Earth's gravity, however, we would fly off into space.

The strength of gravity increases proportional to the masses involved and decreases with the square of their distance apart. If gravity had been slightly weaker by the smallest degree at the moment of creation, it would not have been able to pull matter together to form stars and planets. If it had been slightly stronger to the same degree, it would have pulled matter back into a big crunch before stars and planets were able to form. The extraordinary fine-tuning of gravity is explained by physicist Robin Collins, who asks us to imagine a gravity dial broken down into one-inch increments that stretches right across the universe. He notes that if we moved gravity's setting just one inch out of those unimaginable trillions from its current setting, it would increase gravity by a billion-fold and crush the universe into a super-dense mass (in Strobel, 2004, p. 161).

On the other hand, if the dial was moved in the other direction reducing gravity by as little as five percent, the gravitational force holding the Earth's interior would be weaker. That would result in worldwide earthquakes, volcanos, and tsunamis, cloaking the Earth with layers of volcanic goo. The weaker gravitational pull of the Sun on the Earth would move the Earth to a more distant orbit, resulting in less heat reaching the Earth. The Sun would also cool down since it needs the crush of gravity to push hydrogen atoms toward its core where they fuse together and release heat energy. The combination of the accumulation of ice and ash resulting from all these events would eventually turn the Earth into a giant snowball and us into ice statues. We can thank God that the anthropic gravity dial is set just right.

Electromagnetism

The electromagnetic force is the interaction between electrically charged particles and is carried by quanta of light called virtual photons traveling at the speed of light. It is the combination of all electrical and magnetic forces and is the best understood of the fundamental forces. We depend on electromagnetic energy when we watch TV, make a phone call, or employ any number of other electronic gizmos. Electromagnetic energy travels in waves in a broad spectrum from very long radio waves, which are harmless, to very short gamma rays, which are deadly. Visible light is only a small portion of this spectrum. Our eyes can only detect this part of the spectrum, but various other devices such as radios, x-ray machines, and PET scanners detect other portions. The Sun is a source of energy across the full spectrum. The Earth's atmospheric gases such as ozone and carbon dioxide, as well as its magnetic shield, protects us from exposure to the range of harmful higher energy waves such as gamma rays, x-rays, and most ultraviolet waves.

The electromagnetic force makes chemical bonding possible and gives matter its strength, shape, and hardness. Just as gravity holds everything together on a cosmic scale, the electromagnetic force holds everything else together. If electromagnetic bonding in the nuclei was the slightest bit weaker, electrons could not be held in orbit, and if it was slightly stronger electrons could not bond with the electrons of other atoms to make the molecules and compounds that make us. The electromagnetic force, like gravity, has an infinite range, with its strength proportional to the inverse square of the distance. Unlike gravity which only attracts, the electromagnetic force both attracts and repulses (as in magnets with like poles being pushed together). The electromagnetic force is so powerful that in comparison the contribution of other fundamental forces as determiners of atomic and molecular structures is negligible.

As small as the relative contributions of the other forces are, they are still very important because without them the electromagnetic force would be useless. Paul Davies notes that if the ratio of the strong force to the electromagnetic force had been different by 1 part in 10^{16} the stars could not have formed. Davies also tells us that if the ratio of the electromagnetic force to the gravitational force were increased by one part in 10^{40} only small stars can exist, and if it were decreased by the same amount there would be only large stars. "You must have both large and small stars in the universe: the large ones produce elements in their thermonuclear furnaces; and it is only the small ones that burn long enough to sustain a planet with life" (in Lennox, 2009, p. 70).

The Strong Force

The strong nuclear force is by far the most powerful of the four forces, but it has the shortest interaction distance because it is confined to the nuclei of atoms. The strong force is carried by the gluon boson, so-called because gluons "glue" the protons and neutrons together in the atom's nuclei. Each atom is made up of a number of positively charged protons, and as we know, positively charged objects brought close together will repel one another by the action of electromagnetic force. Despite this repulsion, protons must have a way of sticking together or we would have no elements heavier than hydrogen or helium, and thus no life. It is the strong force that overcomes the proton's natural tendency not to bond with others and explains why atomic nuclei do not fly apart. The strong nuclear force is also the force that powers the stars by crushing hydrogen atoms so tightly that their nuclei overcome their natural repulsion and fuse together, which results in the massive energy that keeps the stars alive.

The mass of an atom's nucleus is slightly *less* than the sum of the masses of its constituent protons and neutrons (collectively called nucleons). This phenomenon exists because when protons and neutrons come together to form a nucleus, a small portion of their mass is converted to energy (recall Einstein's mass/energy equivalence formula: $E = MC^2$). Martin Rees informs us that the mass converted to energy is .007 of the nucleon's initial mass, but if it was .006, a proton would not bond to a neutron to make helium and the universe would consist only of hydrogen. On the other hand, if it was .008, there would be ready and rapid fusion, and no hydrogen would have survived (in Lemley, 2000, p.64).

Elements like to have an equal number of protons and neutrons because if they don't, their binding energy is not strong enough to hold the nucleus together. Additional neutrons upset the binding energy and cause the atom to become unstable. An unstable atom wants to get back to a balanced state and does so by shedding either charged particles or electromagnetic rays (radiation) depending on the nature of its instability. All elements with atomic numbers greater than 83 (starting with polonium) are radioisotopes, meaning that these elements have highly unstable nuclei and are radioactive. For instance, the atomic number of uranium is 92, meaning that it has 92 protons and 92 electrons. The mass number of the isotope is 238, so it has 238 - 92 = 146 neutrons. It is the 54 excess neutrons that make uranium 238 highly radioactive.

The Weak Force

While gravity, electromagnetism, and the strong nuclear force hold things together, the weak nuclear force helps to make things within atoms come apart by radioactive or nuclear decay. The decay caused by the weak force is vital for building different elements. During what is called beta decay, a neutron is replaced by a proton or a proton by a neutron, with an electron being ejected from the nucleus. This interaction between subatomic particles is what we call the weak force and is carried by the W boson (W after the Weak force), which has either a positive or a negative charge, and Z (so-called because it has Zero charge). The stars could not exist without this radioactive decay process. It is this force that drives the fusion of hydrogen protons and neutrons to form deuterium (a rare isotope of hydrogen with a nucleus consisting of one proton and one neutron). The energy generated from nuclear fusion is the source of the heat we get from the Sun. The tiniest increase in the strength of the weak force would drive the hydrogen-to-deuterium process faster, making stars use up their energy faster than their planets could cool, and thus life could not have developed on Earth. A weaker force may have been too feeble to do much fusing at all, and all we may have in the universe is hydrogen.

As weak as it is, the weak force plays a crucial role in life. The heavier elements necessary for life are formed in giant stars and spewed into space in supernovae explosions. Supernovae explosions fuel the cosmic cycle by pollinating the new stars formed from its gasses and dust containing the heavy elements. Such explosions would not occur if the weak force was not exquisitely calibrated. As Paul Davies explains: "If the weak interactions were slightly weaker, the neutrinos [neutrinos are similar to electrons, but they do not carry an electric charge] would not be able to exert enough pressure on the outer envelope of the star to cause the supernova explosion. On the other hand, if it were slightly stronger, the neutrinos would be trapped inside the core, and rendered impotent" (Davies, 1982, p. 68).

The fine-tuning of the relationships among subatomic particles is noted by Hawking and Mlodinow: "If protons were 0.2 percent heavier, they would decay into neutrons, destabilizing atoms. If the sum of the types of quark that make up a proton were changed by as little as 10 percent, there would be far fewer of the stable atom nuclei of which we are made; in fact, the summed quark masses seem roughly optimized for the existence of the largest number of stable nuclei" (Hawking and Mlodinow, 2010, p. 160). Additional fine-tuning involves the proton-to-electron mass ratio. The mass of a neutron is slightly more than the combined masses of a proton, an electron, and a neutrino. If neutrons were less massive by even the slightest amount, they could not decay without energy input. "If its mass were lower by 1%, then isolated protons would decay instead of neutrons, and very few atoms heavier than lithium could form" (Borwein, and Bailey, 2014, np).

We, as well as everything else in the universe, are compounds of molecules composed of two or more elements. Elements are atoms; atoms are composed of protons .... and so on until at each lower level, the solidity of matter fades away into the vibrations of little strings of energy and statistical equations. What is behind those energy vibrations that seem to be the rock bottom of natural reality? We saw in Chapter two that many great physicists, Nobel laureates included, think of the universe as a great thought, and that reminds us once again of John1.1: "In the beginning was the Word, and the Word was with God, and the Word was God."

CHAPTER FIVE

Finding God in the Macro World: The Cosmos

The Big Bang

Who or what created the universe? The scientific answer up to the mid-twentieth century was that it had no beginning; it had always existed. A universe that was beginningless, static, eternal in time, and infinite in space and matter, was accepted as simply a brute fact. This view both relieved scientists of getting into messy metaphysical questions about what caused the universe to exist. The Christian view is that: "In the beginning, God created the heavens and the earth." God's creation was *creatio ex nihilo* (creation from nothing), which was deemed absurd by scientists since nothing (no-thing) comes from nothing. The eternal steady-state model of the universe was so entrenched that when Einstein formulated his general theory of relativity, he was unsettled to find that his theory predicted an expanding universe. He "corrected" his equations by adding a value, later known as the "cosmological constant," to represent a repulsive force countering gravity's attraction, and thus leaving the universe static. He called this the greatest blunder of his life because his initial equations turned out to be right—the universe had a beginning and is expanding. Dark energy is now considered the repulsive force countering the attraction of gravity in an extremely finely tuned balancing act.

In the 1920s, Belgian priest and physicist, Georges Lemaitre, took Einstein's equations, removed the fudge factor, and saw that they supported the model of an expanding universe. Lemaitre reasoned that a past eternal universe would have long ago pulled all the matter together into one huge mass. To avoid this, the universe had to be expanding, and if it was expanding, it had to do so from a finite point in space and time. Lemaitre also reasoned that the expansion force must slightly exceed the gravitational force and that by rewinding the cosmic tape we should arrive at a point when all matter was condensed into a single entity, which he called the "primeval atom." Today, physicists call this the singularity; a "point" of infinite density and infinite temperature, yet zero volume. Everything that exists in the universe, every last atom of matter, every physical force, and spacetime itself was contained in this dense concentration of energy. The singularity was not some tiny dot hanging around somewhere in space because there was no "somewhere" for it to be, nor was it hanging around waiting to pop into existence because there was no time before it popped; it popped on a day without a yesterday.

In 1929-1930, astronomer Edwin Hubble provided evidence for Lemaitre's expanding universe. Hubble's observations of the cosmos showed that all galaxies are moving away from us and away from each other, and that the farther away they were the faster they were moving. This was determined by examining the wavelength spectrum of stars, with galaxies farther from us being the reddest (more "red-shifted") as the light wavelength is stretched. This effect is known as the Doppler Effect and is seen in all physical wavelengths. We experience it most clearly with sound, but it is equally true of the color of light. We experience it every time we hear the sirens of emergency vehicles. As they come closer to us the sound waves are compressed, and as they recede, they are stretched as the siren's pitch decreases. The inescapable conclusion from these observations was that some gigantic event caused the universe to expand with unfathomable force some 13.8 billion years ago, give or take a few million years.

This creation event became known as the Big Bang. The Big Bang brought all matter/energy, space, and time into being in a split-second flash. It was not an expansion into previously unoccupied space like a balloon being blown up, because unlike balloons the universe has no center or edges to expand into. The Big Bang created space as it expanded, and so it was not an explosion *in* space but an explosion *of* space. The attractive force of gravity pulling matter back in had to be exquisitely calibrated to the "explosive" force driving it forward. How exquisite was this balance? Physicist Paul Davies informs us that if the rate of expansion from the beginning differed by more than 10^{-18} seconds we wouldn't be here: "The explosive vigour of the universe is thus matched with almost unbelievable accuracy to its gravitational power. The big bang was not evidently, any old bang, but an explosion of exquisitely arranged magnitude" (Davies, 1984, p. 184).

The Kalam Cosmological Argument for the Beginning of the Universe

While scientists at the time had been reluctant to accept a beginning of the universe, Christians have always known it had a beginning. Astronomer Robert Jastrow notes how science caught up with theology on the matter of creation: "For the scientist who has lived by his faith in the power of reason, the story ends like a bad dream. He has scaled the mountain of ignorance; he is about to conquer the highest peak; as he pulls himself over the final rock, he is greeted by a band of theologians who have been sitting there for centuries" (1992, p. 107). A logical argument for a beginning is the Kalam cosmological argument. It relies on deductive logic such that if the major and minor premises are accepted as true, the conclusion must be accepted. The argument asserts that the universe had a beginning using the logic of cause and effect. Philosopher William Craig (2010) puts the Kalam argument in the form of the following syllogism:

1. *Whatever begins to exist has a cause.* The major premise is self-evident; every physical object or system, or change in that object or system, has a cause preceding the effect. It could not be infinitely old because physicists now realize that infinity is not physically realizable. Stephen Hawking noted that the notion of an infinite universe runs into the brick wall of the second law of Thermodynamics, and states that there had to have been a beginning: "Otherwise, the universe would be in a state of complete disorder by now, and everything would be at the same temperature. In an infinite and everlasting universe, every line of sight would end on the surface of a star. This would mean that the night sky would have been as bright as the surface of the Sun" (Stephen Hawking website, nd). In other words, we know that in some finite time in the future the universe will reach maximum entropy, so if there was an infinite past, we would already be there.

2. *The universe began to exist.* While the minor premise logically follows from the major premise, we have seen that the standard assumption prior to evidence of the Big Bang was that the universe was past eternal. A beginning of the universe made many scientists uncomfortable because it suggested that something caused it to come into being since nothing can be the cause of itself. To do so, the cause would have to have existed prior to itself, which is absurd.

3. *The universe has a cause*. It is the conclusion of the syllogism that rattled atheist scientists because if the universe began to exist at some finite time in the past, it had to have a cause, or it caused itself (some have seriously advanced this notion) or it was caused by a supremely intelligent being. Craig provides some descriptors of this being: "A cause of space and time must be uncaused, beginningless, timeless, spaceless, immaterial personal being endowed with freedom of will and enormous power. And that is a core concept of God" (in Strobel, 2004, p. 132).

Early Big Bang Opposition

As noted, most scientists in the early twentieth century could not fathom a universe that had a beginning and continued to deny that it had. Georges Politzer denied the Big Bang because it violated his materialism: "The universe was not a created object. If it were, then it would have to be created instantaneously by God and brought into existence from nothing. To admit Creation, one has to admit, in the first place, the existence of a moment when the universe did not exist, and that something came out of nothingness. This is something to which science cannot accede" (in Yahya, 1999, p. 19). Biologist John Maddox called the

Big Bang "philosophically unacceptable" because: "Creationists and those of similar persuasions seeking support for their opinions have ample justification in the doctrine of the Big Bang" (Maddox, 1989, p. 425). Robert Jastrow points out why there was such opposition to the Big Bang: "This religious faith of the scientist [in ontological materialism] is violated by the discovery that the world had a beginning under conditions in which the known laws of physics are not valid, and as a product of forces or circumstances we cannot discover. When that happens, the scientist has lost control. If he really examined the implications, he would be traumatized" (Jastrow, 1981, p. 19).

Many scientists were traumatized, and even the phrase "Big Bang" was a cynical coined by Fred Hoyle, who opined that: "The reason why scientists like the 'big bang' is because they are overshadowed by the Book of Genesis. It is deep within the psyche of most scientists to believe in the first page of Genesis" (in Wallace, 2016, p. 101). Surely it is more likely that the Big Bang influenced scientists to accept Genesis than the other way around. Astronomer Allan Sandage considered the "Grand old man of cosmology," initially said of the Big Bang: "It is such a strange conclusion....it cannot really be true." Sandage later became a Christian, noting that "It was my science that drove me to the conclusion that the world is much more complicated than can be explained by science. It was only through the supernatural that I can understand the mystery of existence" (in Strobel, 2004, p. 84).

As the Big Bang became mainstream, Nobel Prize-winning physicist Arno Penzias stated: "The best data we have (concerning the big bang) is exactly what I would have predicted, had I nothing to go on but the five books of Moses, the Psalms, the Bible as a whole" (in Schaefer, 2003, p. 49). Jastrow sees the Genesis account of the beginning and the Big Bang as the same thing described in two different languages: "Now we see how the astronomical evidence supports the Biblical view of the origin of the world. The details differ, but the essential elements in the astronomical and Biblical accounts of Genesis are the same: the chain of events leading to man commenced suddenly and sharply at a definite moment in time, in a flash of light and energy" (Jastrow, 1981, p. 19).

The Cosmic Microwave Background Radiation and the Abundance of Light Elements

Hubble's observations of an expanding universe did not conclusively convince all scientists to accept the Big Bang. The discovery of the cosmic microwave background (CMB) radiation was additional evidence that drove remaining naysayers into the camp. The CMB was discovered accidentally in 1964 by Penzias and Wilson working at Bell laboratories working to detect and measure radio waves. To do this they had to eliminate all interference from their receivers, such as radar or broadcasting signals. After weeks of doing everything possible to eliminate the interference, they found an annoying hiss that was coming from every direction of the sky with equal strength. Penzias and Wilson finally concluded that radiation was coming from outside our own galaxy, but they could not explain it.

The CMB had been predicted since the 1940s as the radiation remnant of the Big Bang. The CMB was created at a time in the universe's history called the Recombination Era when the universe had cooled to about 5,000 degrees Fahrenheit. This was cool enough for electrons and protons to "recombine" into hydrogen atoms and to release the CMB radiation. Because the radiation was coming from everywhere at once, it was realized that "Not only was it a real signal, it was evidence for the big bang itself" (Trefil and Hazen, 2007, p. 318). According to NASA scientists, the fact that CMB radiation is detected everywhere we look and has a uniform temperature (-454.765 Fahrenheit) to better than one part in a thousand "is one compelling reason to interpret the radiation as remnant heat from the Big Bang; it would be very difficult to imagine a local source of radiation that was this uniform. In fact, many scientists have tried to devise alternative explanations for the source of this radiation, but none have succeeded" (National Aeronautics and Space Administration; Tests of Big Bang, nd). The hissing and "snow" you see on your out-of-tune television is the microwave echo of creation.

The last piece of evidence for the Big Bang is the abundance of light elements in the universe. Scientists trace the history of the universe all the way back to "Planck time," which is an astounding 10^{-43} seconds after the Big Bang. At this time, the four fundamental forces of nature were one. In the beginning, the universe was so hot (about 80 million trillion, degrees Fahrenheit) that no atoms could form. At 10^{-5} seconds, quarks, the first "matter," appeared. Quarks and antiquarks zipped around in unbound states. They then suddenly formed themselves into threes to form protons and neutrons with electrical charges set precisely to the level required to later capture the electrons needed to form atoms of hydrogen and helium, the light elements. Protons and neutrons have mirror images of themselves called antimatter with the same mass. Matter has a negatively charged electron and antimatter has a positively charged positron. As matter and antimatter particles whizzed around in the hellish maelstrom, they annihilated each other in a flash of radiation, with new particles of both kinds spontaneously arising from the same radiation. Allen and Lidström inform us that: "If the Standard Model were strictly obeyed, there should have been an essentially complete annihilation of matter and antimatter in the early Universe, leaving only photons." They called this a fundamental problem and "an extreme and unnatural fine-tuning in the initial state of the Universe" (Allen and Lidström, 2016, p.10).

At about three minutes, protons and neutrons were able to form stable nuclei. Hundreds of thousands of years later in the Recombination Era the temperature was cool enough that an electron could attach itself to a proton and neutron to form hydrogen and helium atoms. The abundance of the light elements found today is consistent with their creation in Big Bang nucleosynthesis. Deuterium (an isotope of hydrogen) is particularly important because it is much more abundant than could have been produced by the stars. Stars destroy deuterium, which means that its synthesis could only occur in the Big Bang. The 75/23 hydrogen to helium ratio is thought to be the ratio that existed when deuteron, a particle consisting of a proton and a neutron, which as an atom is deuterium, became stable, thus halting the decay of free neutrons with the expansion and cooling of the universe. Heavier elements had to wait for the formation of stars from hydrogen and helium gases because they require the extreme temperatures and pressures found within stars and are cooked up in the process of stellar nucleosynthesis. This process produces elements up to iron; all elements heavier than iron are formed in the massive energy released by supernovae explosions in the process of supernova nucleosynthesis.

Figure 5.1. The Timeline for the Expansion of Space from the Big Bang

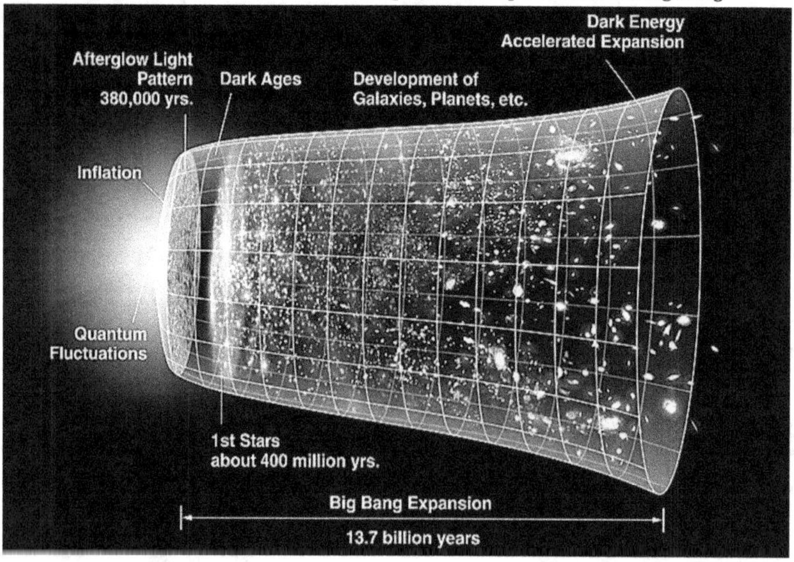

Cosmologists call the time before the formation of hydrogen atoms the cosmological "dark ages" because there was literally no light in the universe. Stars and galaxies provide the universe with light, but they had not yet formed and thus there was no source of visible light. The universe consisted of a dense soup of neutral hydrogen and helium proto-atoms (no electrons). Plenty of photons existed, and light is composed of photons, but at this point in the early universe they ricocheted off free electrons. Photons interact strongly with charged matter and only travel a short distance before being scattered, much as a dense fog scatters the light from a car's headlights. When protons and neutrons were finally able to capture electrons to form atoms, photons were freed to travel through space, leaving behind the CMB radiation. With this decoupling of matter and radiation, we get light, and thus astronomers are able to see back into the universe to about 380,000 years after the Big Bang. Figure 5.1 from The National Aeronautics and Space Administration (NASA) is the timeline for the expansion of space.

Lighting the Big Bang Touch Paper

The most remarkable fine-tuning of all is getting the universe started in the first place. From a theological point of view, God simply spoke the universe into existence; from a scientific point of view, physicists have found the fingerprints of God's creation in the "Creator's aim" in phase-space. Phase-space is dynamic multidimensional space in which all possible coordinates are represented, with each coordinate specifying the possible state of a physical system. The state of the universe at the Big Bang is intimately connected to the second law of thermodynamics and entropy. Recall that entropy is the degree of thermodynamic disorder, which in a closed system is always increasing. Given this, there had to be an immense degree of order at the Big Bang because a universe capable of producing intelligent life must begin with the lowest possible degree of entropy—the possibility of a high entropy universe is immeasurably greater.

Nobel laureate Sir Roger Penrose asks us to imagine all the possible ways that the universe might have started off in phase-space and the probability that the Creator could hit the exact point to create a life-producing universe. He calculated the probability of the initial entropy conditions of the Big Bang by calculating the maximum entropy of the universe. This figure is an estimate of the total phase-space volume available to the Creator and is the logarithm of the total phase-space volume of all possible beginnings of the universe, or 10^{123}. Logarithms and exponents are inverse functions, so the total phase-space volume is $10^{10^{(123)}}$. Penrose asks: "How big was the original phase-space volume W [W = original phase-space volume] that the Creator had to aim for in order to provide a universe compatible with the second law of thermodynamics and with what we now observe?" He then remarks on the two ways to estimate this figure and writes: "Either way, the ratio of V [total phase-space volume] to W will be, closely V/W = $10^{10^{(123)}}$" (2016, pp. 445-446). Penrose states that this huge number could not be written down if we had every elementary particle in the universe to write a zero on. To put it another way, the probability is vastly smaller than the probability of one person winning the Powerball jackpot every day for the 13.8 billion years the universe has existed, with everyone on Earth buying a ticket.

Penrose's calculations present problems for physicists who think only in terms of materialism. For instance, three eminent physicists addressed the issue stating: "The question then is whether the origin of the universe can be a naturally occurring fluctuation, or must it be due to an external agent which starts the system out in a specific low entropy state?" (Dyson, Kleban, and Susskind, 2002, p. 3). They note that it is an undisputed fact that the universe only makes sense if it began in a state of minimal entropy, and added: "there is no universally accepted explanation of how the universe got into such a special state. In this paper, we would like to sharpen the question by making two assumptions which we feel are well motivated by observation and recent theory. Far from providing a solution to the problem, we will be led to a disturbing crisis." The "disturbing crisis" is nothing less than forcing materialist cosmologists to think the unthinkable: "Another possibility is an unknown agent intervened in the evolution, and for reasons of its own restarted the universe in the state of low entropy characterizing inflation" (Dyson, Kleban, and Susskind, 2002, p. 1). There is no reason outside of God himself for His creative acts. God created a universe susceptible to scientific analysis so that His creatures can glimpse his awesome power.

The Geography of the Universe and the Cosmological Constant

After the period of recombination, the universe was a homogeneous mass of matter and energy of immense density. For matter to coalesce into galaxies there must be some contrast or "roughness" in the smooth homogeneity of the distribution of matter to enable it to collapse under the pull of gravity. The formation of galaxies thus depends crucially on matter density variation from one location to another, but this variation must be very small. We can see these density perturbations in the images supplied by the Wilkinson Microwave Anisotropy Probe, a mission charged with measuring temperature differences in the CMB across the visible universe. "Anisotropy" refers to small temperature fluctuations in the background radiation, and is the opposite of isotropy, or universal homogeneity. Scientists note that anisotropy is fine-tuned to about one part in 100,000. If it had been "significantly smaller, the early universe would have been too smooth for stars and galaxies to have formed" ... and "galaxies would have been denser, resulting in numerous stellar collisions, so that stable, long-lived stars with planetary systems would have been very rare" (Bailey, 2018, np). In other words, the *contrast* and *density* of matter had to be anthropically fine-tuned from the very instant of the Big Bang.

There is a critical value (p_{crit}) of energy density that prevents gravity from overcoming the force of expansion and pulling all matter into a big crunch that had to vary by less than one part in 10^{60} from the very beginning of creation. Paul Davies says of this fine-tuning: "We know of no reason why p is not a purely arbitrary number...to choose p so close to p_{crit}, fine-tuned to such stunning accuracy, is surely one of the great mysteries of cosmology" (Davies, 1982, p. 90). NASA scientists tell us that: "The value of the critical density is very small: it corresponds to roughly 6 hydrogen atoms per cubic meter [that's just over 35 cubic feet], an astonishingly good vacuum by terrestrial standards!" (National Aeronautics and Space Administration, 2019, np).

The density of matter in the universe affects the geometry of space-time, with the critical value to fit the requirement for a flat universe. Only in a flat universe is the energy of matter balanced by the energy of the gravity mass creates. When physicists say the universe is flat, they mean that their data suggest that the geometry of the universe has no curvature. That is, the geometry of the universe is such that parallel lines will never cross, the angles in a triangle will always add up to 180 degrees. There is a lot of insanely difficult mathematics used to describe this geometry, but suffice to say that the geometry of the universe is determined by its relative density and that the CMB and the distribution of galaxies comprise the measuring rods. In a spherical universe with greater than critical density, Euclidean geometry breaks down; the three angles of a triangle no longer equal 180 degrees, and lines that start out parallel will eventually meet. A universe with such density will curve space-time in on itself and gravity will collapse it in a "big crunch." This was the basis for the now-discredited oscillating universe of endless cycles of big bangs and big crunches. In a saddle-shaped (hyperbolic) universe, there is insufficient mass to cause the expansion of the universe to stop and it will expand forever. Our flat universe will also expand forever, but with the rate of expansion gradually approaching zero after an almost infinite amount of time.

In the early universe, matter was clumped much closer together, and therefore there was greater mass density straining against gravity's grip. Initially, the repulsive force of the Big Bang was enough to balance this out, but this has to dissipate eventually, thus requiring another force to prevent a "big crunch." The Supernova Cosmology Project began in 1998 expecting to measure the deceleration of the universe but found that it was accelerating. Gravity needed to dominate during the period of matter accretion into galaxies, stars, and planets, but for some reason known only to God, dark energy now rules the roost. Einstein's cosmological constant is this dark energy built into the vacuum of space and is the force keeping the universe expanding. There is much about the amazing precision of the cosmological constant that puzzles physicists. Jenkins and Perez remark that the "most serious fine-tuning problem in theoretical physics: the smallness of the 'cosmological constant,' thanks to which our universe neither recollapsed into nothingness a fraction of a second after the big bang, nor was ripped apart by an exponentially accelerating expansion" (Jenkins and Perez, 2010, p. 44).

Physicists Livio and Rees argue that anthropic reasoning is seriously discussed in physics for sorting out certain cosmological phenomena, such as the mystery of the cosmological constant. They ask: "Why is the force so small? If there was an inflationary era with a large cosmic repulsion, how could that force have been switched off (or somehow have been neutralized) with such amazing precision? In our present universe, Λ [Λ "lambda," the symbol for the cosmological constant] is lower by a factor of about 10^{120} than the value that seems natural to theorists" (Livio and Rees, 2005, p. 1022). The "switching off" or "neutralization" of repulsion they refer to must be exquisitely fine-tuned to 120 decimal places from the moment of creation. They also note that: "If Λ were larger, then the acceleration would have overwhelmed gravity before galaxies had a chance to form" (Livio and Rees, 2005, p. 1022). Nobel laureate Steven Weinberg's understanding of the razor's edge balance between dark energy and gravity lead him to exclaim: "This is the one fine-tuning that seems to be extreme, far beyond what you could imagine just having to accept as a mere accident" (in Folger, 2008, np). Even Leonard Susskind, who hates the anthropic principle, acknowledges its power: "The fact that [the cosmological constant] is not absent is a cataclysm for physicists, and the only way that we know how to make any sense of it is through the reviled and despised Anthropic Principle" (Susskind, 2005, p. 22). How utterly strange that the only thing Susskind says makes sense of the cosmological constant is something he considers "reviled and despised." Recall that the Big Bang was also reviled and despised by previous generations of atheist scientists.

Chapter Six
Our Cosmic Neighborhood

The Milky Way

If the Earth is our house and the solar system our neighborhood, the Milky Way galaxy is our city in a universe of cities. Of all the possible billions of galaxies in the vastness of space, there is no better galaxy than the Milky Way in which to reside. The Milky Way is in the top one or two percent of galaxies massive enough to have sufficient building material to construct such a special planet as Earth. Large galactic mass means that it has gravity strong enough to attract sufficient galactic gases (mostly hydrogen) to construct the heavier elements we need for life, such as carbon, oxygen, nitrogen, and phosphorus. Hydrogen is transformed in the stars into these building blocks of life when the unimaginably immense heat and pressure in a star's core cause fusion reactions. The primeval fusion is two light hydrogen nuclei merging to form a single heavier helium nucleus. The resulting helium nucleus mass is less than the sum of the mass of the two nuclei that made it. That mass is lost in the form of energy (because of the equivalence of mass and energy).

In the first few minutes of the universe's existence, it was composed of a swirling mass of protons and neutrons not yet combined with electrons. These were the primordial building blocks of everything, including us. Eventually, we got hydrogen and helium, but you can't build a planet from these light elements; you need the heavy elements, and these take time to be fabricated in the stars in stellar nucleosynthesis. Stars are in a perpetual battle between gravity and nuclear fusion. Gravity tries to squeeze the star into the smallest possible ball of matter, but the burning of the nuclear fuel in the star's core creates strong countering outward pressure against the inward squeeze of gravity. The more a star is crushed by gravity, the hotter and denser it gets, and the faster the rate of nuclear fusion. As long as the star has fuel, its fusion energy will push back against the crush. The most massive stars burn their fuel supply quickly to provide the energy needed to counteract their massive gravitational crush and may go supernova after only 3 to 10 million years. A star with our Sun's mass can keep on fusing its hydrogen for about 10 billion years.

All stars are born with a finite amount of hydrogen that will eventually run out. The older stars in the universe have burned long enough to produce many of the heavier elements, but the heavier the element the less the fusion energy produced. The most massive stars (about eight times larger than the Sun) fuse progressively heavier elements (larger numbers of protons, neutrons, and electrons) until it produces iron, which cannot be burned further. At this point, the energy required to fuse iron is more than the energy the star gets back in return. With no energy being released, the star's core collapses in a matter of seconds under the tremendous crush of gravity. The collapse happens so quickly that it creates tremendous shock waves that cause the outer part of the star to explode—to go supernova and spew its elements out into space. The original progenitor star either collapses to a neutron star or black hole or is completely destroyed. Of course, the particulars of supernovae events are very complicated, and there are different types of supernovae depending on the type of progenitor star or whether the progenitors are binary stars orbiting each other.

But iron is only the 26th element in the periodic table, so how do we get the other 66 found naturally on Earth? The elements past iron are forged by the mighty energy of supernovae explosions. When a star is about to go supernova, the heat of the core increases so quickly that iron is smashed apart into helium, which, in turn, is shattered into protons and neutrons. The tremendous energy released by a supernova releases a superabundance of free subatomic particles cascading from the collapsing core. This huge heat energy drives massive fusion reactions among these particles that forge the remaining elements found on Earth. Your cosmic city and neighborhood got built because stars died and gave us the materials a few billion years ago. However, enough of them had to die close enough, and at just the right time, to have the elements captured in the dust and gases of the Sun's nebular (see below) when it was forming. If that kind of supernovae activity occurred in our galactic neighborhood today, the radiation from them would exterminate all life on Earth. We can thus add exquisite timing to our long list of anthropic "coincidences."

The Milky Way galaxy is a spiral galaxy. For a number of reasons, only a spiral galaxy such as ours is capable of sustaining intelligent life; elliptical and irregular galaxies are not. Elliptical galaxies are spherical or egg-shaped and contain mostly ancient stars with limited resources for building planets. Star formation ceases too early in elliptical galaxies to produce sufficient amounts of the heavy elements, and the stars they contain are too crowded to sustain stable planetary orbits. The situation is even worse in irregular galaxies, so-called because they come in a variety of shapes. Irregular galaxies contain no spiral arms or nuclear bulges, and their stars have chaotic orbits which take them periodically into the vicinity of planets, disrupting their orbits and bathing them in deadly radiation. Our own Milky Way also has many unsavory neighborhoods containing densely packed collections of millions of ancient stars revolving around the galactic core called globular clusters. Earth could not exist in a globular cluster because its ancient stars are poor in the heavier elements necessary to build such a planet. The gravitational pull of the myriad of stars in a globular cluster results in highly elliptical orbits of its planets that either plunges them into each other or into extremes of heat and cold as they move closer or farther from their stars (Ross, 2016).

The Galactic Habitable Zone

The Milky Way galaxy is about 180,000 light-years across. Our zone of the galaxy is safely tucked away about 27,000 light-years from its center, and tens of thousands of light-years away from the outer rim of the galaxy's spiral arm. Remember, a light year is the distance light travels in a year at 186,282 miles per second, so one light-year is equal to 5,878,625,370,000 miles. Multiply that by 27,000 and you will get a *very* large number. Our solar system orbits Sagittarius A* (Sagittarius A star) about once every 200 to 250 million years traveling at about 500,000 miles an hour. Despite its name, Sagittarius A* star is not a star; it is rather the matter (over 4 million times more mass than our Sun) revolving around the supermassive black hole at the center of the galaxy.

Our neighborhood in the galaxy is in what is known as the Galactic Habitable Zone (GHZ). It is a habitable zone because it is a band of space inside the galaxy with physical conditions that are compatible with the building of a resource-rich planet, and the origin and development of life. As Gonzales, Brownlee, and Ward tell us: "The boundaries of the galactic habitable zone are set by two requirements: the availability of material to build a habitable planet and adequate seclusion from cosmic threats" (Gonzales, Brownlee, and Ward, 2001, p. 62). The center of the Milky Way is a very dangerous place full of exploding supernovae and a gigantic black hole, but the GHZ band is far enough from the center to avoid the effects of deadly radiation of exploding stars or the possibility of drifting too close to the black hole and getting sucked in. On the other hand, our solar system is close enough to benefit from the heavy elements that supernovae explosions spew out into space. If it were any further out on the fringes of the spiral arms of the galaxy, there would not be enough heavy elements to build Earth-like planets and we would be exposed to hazardous giant gas clouds the spiral arms often visit. As astrophysicists Hugh Ross and Guillermo Gonzalez explain:

> The solar system occupies a position in the disk of the Milky Way approximately halfway to its edge and in-between two spiral arms. We now know enough about the structure of our galaxy to understand why our location should be preferred over others. If our solar system were closer to the center of the Milky Way or closer to one of its spiral arms, we would encounter harmful radiation from supernovae and perturbations from stars that would send Oort cloud comets careening into the inner

solar system. If the solar system had formed farther out in the disk of the Milky Way, there would not have been sufficient heavy elements to build a planet capable of supporting life. (Gonzalez and Ross, 2000, np)

Our solar system exists as part of the galactic spiral called the Orion Arm located between two other spiral arms—Sagittarius and Perseus. Stars in the Orion Arm orbit around the galactic center close to the corotation circle. Being close, but not too close, to the corotation circle, is defined as "the precise distance from the galactic center at which stars rotate around the center at the same rate as the galactic arms." In other words, our solar system moves around the galactic center at the same speed as the spiral arms. Stars closer to the center of the galaxy move faster, and stars farther away move slower. Both are in danger of crossing the spiral arms, which disrupts planetary orbits. Moving at identical speeds means that our Sun will remain between the other two spiral arms and not wander into them. Most other solar systems do not evidence our orbital pattern, and very few of our neighbors enjoy the same level of safety. Only about one in 100,000 stars in a spiral galaxy are in a corotation radii.

Stars at the fringes of spiral arms are too far from the galaxy's "industrial zone" to benefit from supernovae heavy elements necessary for planet formation. Fortunately, our solar system is far away from the spiral arms and orbits the center in an almost perfect circle (very low eccentricity). Most other solar systems do not evidence this orbital pattern, and thus very few of our neighbors enjoy the same level of safety. As Plaxco and Gross write about our "just right" orbit: "Such low eccentricity orbits are, however, relatively, rare, [they tend to be elliptical rather than circular] and the majority of Sun-like stars currently in our neighborhood spend a significant fraction of each galactic orbit far too close to the galactic center for comfort...less than 5% of all stars lie in the life-supporting zone" (Plaxco and Gross, 2006, p. 35). Just because they exist in the life-supporting zone, however, it does not mean that they have life-sustaining planets. A multitude of other physical, chemical, and biological conditions must be exquisitely fine-tuned for the emergence of complex intelligent life.

Only a spiral galaxy can support an Earth-like planet, but not any old spiral galaxy will do either. Such a galaxy can be neither too big nor too small. Galaxies more massive than the Milky Way spawn massive back holes with enough mass to ignite jets of deadly radiation, and their immense gravitation causes chaotic mergers with many smaller galaxies. Both too large and too small galaxies run into the problem of the corotation radius. In larger galaxies, the corotation is far too long for the heavier elements to reach any planets that may otherwise have the potential for advanced life. The problem for smaller galaxies is that the co-rotation radius is too short for life because planets will be too close to the galactic core's deadly radiation. As Gonzales, Brownlee, and Ward so aptly put it: "We live in prime real estate" (2001, p. 67).

Being located in the galaxy's "sweet spot" is only the absolute minimum requirement for a home with intelligent life. A suitable home must also have a suitable star, and must not be too close to it or too far from it. Too close to it, and water boils away; too far and it becomes permanent ice. You then need a planet that is neither too massive nor not massive enough, and with the right amount of internal activity to help create an atmosphere and to provide a protective shield against harmful radiation. We will address these planetary matters in the following chapter. For now, let us take a look at our marvelous "just right" star we call the Sun.

The Sun

The Sun is not just any old star but rather one—and perhaps the only one—that has all the necessary characteristics to make complex life possible. Unlike about 85% of other stars, the Sun is solitary. The other 85% are locked with two or more other stars, which results in wild gravitational pulls that make stable planetary orbits impossible. The Sun is a relatively young star, forming only about 4.6 billion years ago. It formed from an immense cloud of dust and gas called a nebular cloud composed mostly of hydrogen. At some point, the gas cloud began to spiral around its center until gravity overcame the gas pressure and fused hydrogen into helium to ignite our Sun into the flaming ball of plasma it is today. Plasma is often referred to as the fourth state of matter along with solids, liquids, and gas, and is matter superheated to the point that electrons are ripped away from atoms to form ionized gas. The cloud's material not used for the Sun (less than 1%) coalesced into the planets, moons, and other objects in the solar system.

According to NASA scientists, the Sun's diameter is 864,400 miles, which is about 109 times the diameter of Earth and the Sun is 333,000 more massive than Earth. It has a core temperature of about 27 million degrees Fahrenheit and a surface temperature of about 10,000 degrees Fahrenheit. It is composed mostly of hydrogen (about 76%) and helium (about 22%), with traces of other elements such as oxygen and carbon. It is much too cool to be able to produce iron, however. The Sun's energy makes life possible by sending us its precious photons. When atoms are smashed together in the Sun's core, energy is released in the form of photons. In making these photons, the Sun consumes 600 million tons of hydrogen and turns it into 596 million tons of helium every second. The four million tons of mass lost is the energy (photons) produced by fusion. Sunlight is used as energy by plant life to synthesize foods from carbon dioxide and water in the process of photosynthesis. We are thus eating recycled sunlight when we eat plants or eat the meat of animals that live on plants.

Those life-giving photons took just over eight minutes to travel the 92,955,807 miles from the surface of the Sun, but they took a lot longer to travel the

432,168.6 miles from the Sun's core to its surface. Given that they travel at the speed of light—because that's what they are—you might think that it would take them about 2.3 seconds, but you would be way off. This is a good thing, because the photons released when hydrogen nuclei fuse to form helium are deadly gamma rays, and we don't want them hitting us. The gamma rays emitted from the fusion may take thousands of years to leave the core because the core is incredibly dense (only 2% of the Sun's volume but 40% of its mass). Photons move only a microscopic distance before they are absorbed by protons and then re-emitted. Once out of the core, it moves into the Sun's radiative zone. In this super-dense zone (though less dense than the core), photons bounce around wildly as they hit protons and are re-emitted—one step forward, three to the left, one step backward, two to the right, and so on, like a ball in a pinball machine. Depending on the average number of collisions the photon has on its way out, and that depends on the average matter density it encounters, this process can take anywhere from 10,000 to 200,000 years (Trefil and Hazen, 2007). After trillions of collisions, the gamma rays lose energy and become less energetic photons such as those shown in Figure 6.1.

Figure 6.1. The Electromagnetic Spectrum

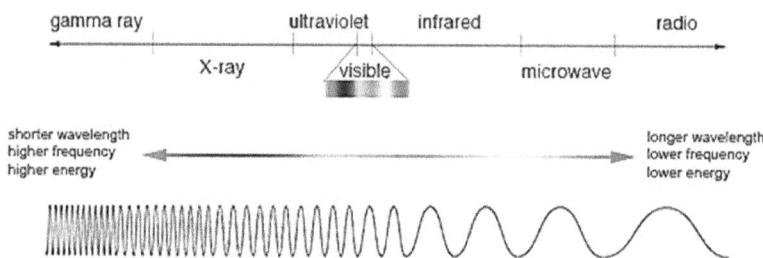

Once these photons escape the radiative zone, they enter the cooler and much less dense convective zone. Here, they are carried toward the surface by hot gas, like bubbles carried upwards in a pan of boiling water. This journey takes only a few days and they can then zip off to Earth as infrared radiation (about 50%; this provides heat), visible light (about 42%), and UV rays (about 8%). Figure 6.1 shows the seven different wavelengths of photons in the electromagnetic spectrum. Each photon contains a certain amount of radiation energy, with each type being progressively less dangerous as we go from short to longer wavelengths.

Further Anthropic Considerations

Stars are categorized according to their mass and luminosity. Luminosity is a star's intrinsic brightness and energy output, which depends on its mass and temperature, and the Sun's brightness and energy are "just right" for life. NASA scientists tell us that the Sun's luminosity is highly stable, varying by only one-tenth of a percent over the last couple of billion years. This is very important because more variety would lead to wild climate changes on Earth. Many stars frequently undergo large increases in luminosity and thus release immense additional heat energy radiating to any bodies orbiting them. We are fortunate that our star is as massive as it is, but it cannot be so massive that it burns out before the Earth acquires water and manufactures an atmosphere. At the other end of the spectrum, low-mass stars are more likely to tidally lock any planets they may have, and their unstable luminosity would sterilize life on an Earth-like planet due to regular large and deadly stellar flares (Barnes, 2017). We are indeed blessed by the Sun's anomalous properties—its mass, luminosity, and stability—all of which make Earth's habitability possible.

As stable as the Sun presently is, before life other than bacteria appeared on Earth, its luminosity was only about 70% of what it is today, meaning that its heat was much less, and the surface of the Earth should have been frozen solid. Ancient rocks indicate liquid water and life existed on Earth at that time, however, so its temperature must have been around what it is today. This paradox used to be explained by a much higher concentration of carbon dioxide and methane in the early atmosphere than today trapping the sparse heat. But that would have required many hundreds of times greater volume of these greenhouse gases, and no such high concentrations are found in ancient soils and sediments, and increasing solar luminosity combined with such an atmosphere would have rendered Earth too hot for habitation. An explanation for why the Earth didn't get this hot is that the oceans were much larger before the emergence of more land, and water is able to trap more heat than land. To date, the problem remains unresolved (Feulner, 2012). However, some anthropic condition(s) must have been present to prevent the early Earth from becoming a frozen planet, or today's Earth (given the about 30% increase in the Sun's luminosity) from becoming an uninhabitable inferno. Can you see God's hand in this?

The Sun sends us so much energy that a lot of it has to be returned to space to keep Earth from overheating. In fact, the Earth sends a little more heat out into space than it receives because the Earth generates its own heat from the radioactive decay that occurs deep inside it. About a third of the Sun's energy is reflected by clouds, water, and snow. What is not reflected is absorbed and used. "How is this balance maintained?" asks a University of California, Davis, online astronomy course on the electromagnetic spectrum, and replies that "Earth warms up to exactly the temperature that is necessary to re-radiate *exactly the right amount of energy*" [my emphasis] (University of California, Davis, nd; np). This is yet another example of the Anthropic Principle at work.

The wind is another source of energy for which the Sun is ultimately responsible. The wind is the movement of atmospheric gases around the planet caused by the Sun's heat. Heat warms the land and bodies of water at different rates because land and water absorb or reflect sunlight differently. This uneven heating results in changes in the atmosphere as hot air rises and cool air moves in to replace it. The flow directly moves around the high/low-pressure systems because of the rotation of the Earth. The Earth's rotation causes the wind to deflect counterclockwise in the northern hemisphere and clockwise in the southern hemisphere (this deflection is called the Coriolis Effect). Although high winds can cause a lot of damage, they play a vital role in keeping the air, land, and sea fresh. It drives the ocean currents and aids plant life to disperse their seeds, spores, and pollen. Wind-driven turbines are also playing an increasing role in providing us with clean and renewable energy. We are indeed blessed. Gonzales and Richards write of the many wonders of our unusual star: "Taken together, then, the anomalies [mass, luminosity, stability] suggest that the Sun is atypical in ways that enhance Earth's habitability for technological life" (Gonzales and Richards, 2004, p.137).

Tides are caused by the gravitational pull of the Sun and Moon. These pulls cause a slight bulge in the Earth that moves the oceans about. Tides clean and oxygenate the oceans and bring vital nutrients from erosion of the earth to them. The tides' currents mix arctic water not able to absorb much solar energy with warmer water from regions that can, which balances planetary temperatures, making for a more predictable and habitable climate. In partnership with the Moon, the Sun generates optimal tides for life on our planet. It does so because while the Sun is about 400 times larger than the Moon, it is also about 400 times further away. That is why their angular radii make it seem as though the Moon and the Sun have exactly the same size as viewed from Earth. The odds against such a perfect match are truly enormous: "Since even a modest difference in the Moon's angular size relative to the Sun's" says astrophysicist Steven Balbus, "would lead to a qualitatively different tidal modulation, the fact that we live on a planet with a Sun and Moon of close apparent size is not entirely coincidental: it may have an anthropic basis" (2014, p. 1). Balbus' article shows how the contrast between the magnitude and timing of the tidal forces the Sun and Moon (the Moon exerts roughly twice the tidal effect of the Sun) exert on the Earth plays an important role in enhancing Earth's biocomplexity, biodiversity, and biomass. The precise calibration of the tides of these heavenly bodies produces numerous coastal tide pools where seafood, a major food source in many cultures, is easily harvested.

Astrophysicist Hugh Ross notes on Balbus's "anthropic basis" for the complex "just right tides that only an all-powerful Designer could be responsible.

The required fine-tuning to get such perfect-for-complex-life tides is such that, in spite of the observable universe containing as many as 50 billion trillion planets, the Sun-Earth-Moon system likely stands alone in generating such optimal tides for its planet. When combined with all the other fine-tuned features of the Sun-Earth-Moon system, nothing short of supernatural, super-intelligent designs comes close to offering a reasonable explanation. (Ross, 2019, np.)

The Sun as Metaphor for God

Because of its many life-giving properties, certain ancient civilizations worshiped the Sun as a god. Of course, the Sun is only a giant ball of hot gas, and unlike God, it had a beginning in time and will eventually die. Yet the Bible is chock-full of metaphorical references to the Sun, the light of our lives. For instance, Psalm 84:11 states: "For the Lord God is a sun and shield." We take this allegorically to mean that but for the love of God that enlightens, guides, and directs us, we humans would walk in darkness. Just as the physical Sun provides light, warmth, and beauty to our bodies, the personhood of God is the source of light, joy, and happiness, for our souls; the light of the world that dispels darkness. God is also a shield of faith against the dark temptations of the flesh. Also like the Sun, God is distant, but we feel his rays like we feel the Sun's. Like the Sun when it shines on the other side of the Earth, God is there even when we can't see him are can't look directly at him. G. K. Chesterton said it best: "God is like the sun, you cannot look at it but without it, you cannot look at anything else." Another great British writer, C. S. Lewis, likewise declared, "I believe in Christianity as I believe that the Sun has risen, not only because I see it but because by it, I see everything else."

CHAPTER SEVEN

Our Very Special Earthly Home

The Circumstellar Habitable Zone

We occupy a very special place in the cosmos called Earth, which Carl Sagan famously dubbed the "pale blue dot" upon viewing pictures of Earth sent from Voyager 1 from 3.6 billion miles away. Earth is referred to as the "Goldilocks" planet because it seems to be the only place in the universe where everything seems just right for intelligent life. Just as our solar system is in a prime real estate area in the universe, so is Earth's location within the solar system. Earth is located in the Circumstellar Habitable Zone (CHZ), which is a band of space around the Sun that is hospitable to life because only planets residing there can hold surface water. The CHZ band is measured in astronomical units (AUs). One AU is the distance between the Earth and the Sun, or 92,955,807 miles = 1.0 AU. A typical estimate of the CHZ is between 0.95 and 1.37 AUs.

Figure 7.1. Habitable Zones Around Different Types of Stars

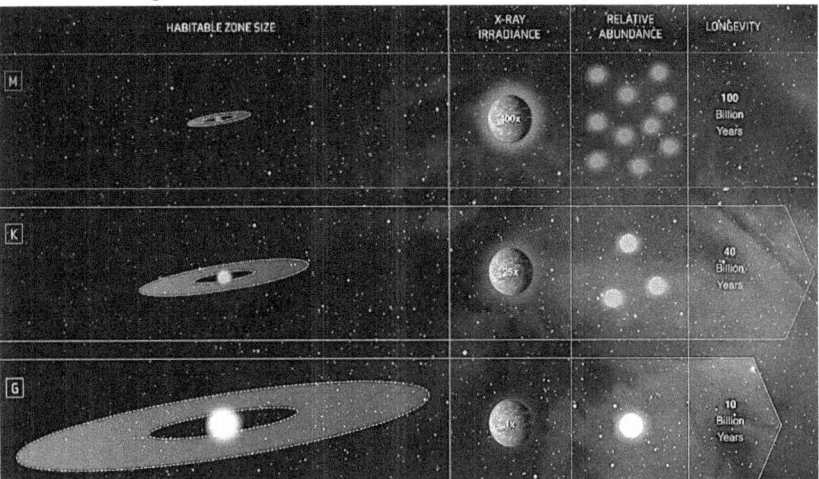

Figure 7.1 presents habitable zones around three different star types: M dwarfs, K dwarfs, and G main-sequence stars. Observe the narrowness of M and K stars' habitable zones and how much x-ray irradiance to which a planet is exposed. Some astronomers see K-star systems as the "sweet spot" for life simply because K dwarfs will burn four times longer than G-stars (a longer time to kick-start life) while not subject to the massive irradiation of M-stars. However,

their 25x greater radiation than experienced around G main sequence stars (it could even be 100x) would mean that their planets would be exposed to life-precluding levels of radiation (National Aeronautics and Space Administration, 2020). Thus, the requirement that a planet has liquid surface water is not enough by itself for habitability.

Our prize patch of cosmic real estate was formed from the materials left over from the creation of the Sun. Solar winds blew away most of the lighter elements (hydrogen and helium) from our planet, leaving behind the needed heavier elements. If Earth had been further away from the Sun, the solar winds would have been too weak to have this effect, and the light elements would have ignited into gaseous planets like Jupiter and Saturn. We were thus in the right place at the right time to capture an atmosphere, an ocean, and a landmass; the minimum requirements for life. Could this be a mere coincidence? Gonzales, Richards, and Ward calculated that the probability of getting an Earth-like planet by chance far exceeds Dempski's probability boundary: "even in a universe with 10^{11} stars per galaxy and 10^{11} galaxies, totaling 10^{22} available attempts, the chances of getting one such system would still be one chance in 10^{158}" (Gonzales, Brownlee and Ward, 2001, p. 62).

The Earth's orbit is very special in terms of orbital eccentricity—the degree to which an orbit departs from circularity. The eccentricity values of planetary orbits range from zero (perfectly circular orbit around its star at constant speed) to 1. Our eccentricity value is a mere 0.0167, which makes our weather patterns stable. The closer a planet gets to an eccentricity of 1, its orbit becomes increasingly longer and will eventually escape the parent body's gravity and fly into deep space. This is so because all bodies of matter in space want to keep moving forward in a straight line, not in circles. Without the Sun pulling us toward it, the Earth would zoom away into the cold depths of space. Orbits are the result of this tug-of-war between planets and their parent body. The parent body's gravity wants to pull an orbiting body in, but the body wants to move straight ahead. The result of a precise balance between the forward motion of a body in space and the pull of gravity on it from another body in space is an orbit.

If we were orbiting much closer to the Sun, we may be caught in a tidal lock. Tidal locking means that a body's rotational period equals its orbital period around its parent body. It occurs because the gravitational pull of a parent body slows rotation until one side always faces the parent body and the other always faces away. Because magnetism depends on motion, the reduction of orbital speed will eventually shut down any magnetic shield a planet may have had. Mercury is tidally locked to the Sun and the Moon to the Earth. This results in an extremely hot surface on the side of a planet facing its star, and extreme cold on the side facing away; both being hostile to complex life. This is a major problem for planets within the habitable zones of stars with low luminosity, such as M and K dwarfs. Such planets have to hug their stars closely to capture their heat, and in doing so they will eventually become tidally locked. Unlike most planets, the Earth's orbit around the Sun describes an almost perfect circle, keeping it in the CHZ permanently. Venus's orbit is more circular than the Earth's, but it is close to being tidally locked; it has such a slow rotation that its day of 243 Earth days is actually longer than its year of 225 Earth days. That is, it rotates on its axis slower than it orbits the Sun. Earth's 24-hour rotation keeps it from being too long in the light and heat of the day or in the dark and cold of the night.

If a planet is to be habitable, its mass must not be too small or too large. Of the thousands of exoplanets discovered so far, just about all are more massive than the Earth. More massive planets have greater gravity, which prevents the formation of mountains and continents. They would likely also have a mostly hydrogen/helium atmosphere with more of its mass composed of water (up to 50%) if within its star's CHZ. Earth's water mass is only 0.06% (mass is a measure of how much matter an object has, while volume is the amount of space an object occupies) of its mass. However, a planet must have sufficient mass to hold an atmosphere and greenhouse gases, without which it wouldn't have surface water. Its surface pressure and temperature must also stay in the same range for billions of years to prevent atmospheric escape into space. Gravity must be sufficiently high to prevent this, but it can't be too high that the planet could not rid itself of the thick hydrogen/helium-rich atmosphere of its early existence.

On occasion, the Sun sends eruptions of energy towards Earth called solar flares, which include life-inhibiting x-rays, and gamma-rays that escaped the radiative zone intact. Our first line of defense against this onslaught is the magnetic shield that surrounds the Earth and extends about 370,000 miles out into space. Some radiation penetrates the shield but is channeled toward the poles and produces beautiful auroras. Such a shield is most important in a planet's early life when its host star is more excitable and throws more of its material out into space and strips planets of their atmosphere. The Magnetic field is generated by a molten iron core at the center of the Earth. This solid inner iron core is two-thirds of the size of the Moon and is as hot as the Sun's surface. The pressure of gravity—which is really crushing at close to 4,000 miles below the Earth's surface—prevents it from becoming liquid, but its surrounding outer 1,243-mile outer core of iron, nickel, and some other metals is kept fluid because of lower gravitational pressure there. Because the Earth is spinning, it causes what is known as the Coriolis force. This force causes swirling whirlpools of liquid iron deep within the Earth, generating electric currents, which in turn produce magnetic fields. Because of the spiraling, the different shields align and combine to produce the magnetic shield.

The Earth's atmosphere also provides protection from deadly solar radiation in the form of the ozone layer, which is concentrated in a layer in the stratosphere, about 10 to 20 miles above the Earth's surface. Ozone (or trioxygen) is formed when UV rays split oxygen atoms and three of them become covalently bonded (bonded by sharing their electrons). UV radiation thus creates the very ozone that scatters other UV rays in a continuing cycle. The amount of ozone in the atmosphere has to be finely tuned. Too much ozone would hinder respiration for large creatures such as ourselves and reduce crop yields; too little would lead to increased UV radiation reaching the Earth and would also damage crops and cause serious health risks problems. Ozone deflects about 97% of UV rays, which is a remarkable level of efficiency. Oncologist Arthur Brown writes of this anthropic level of efficiency: "The Ozone layer is a mighty proof of the Creator's forethought. ... A wall which prevents death to every living thing, just the right thickness, and exactly the right defense, gives every evidence of plan" (in Hick, 1963, p. 25).

When the radiation threat comes from beyond our solar system, the Sun protects us. The threat is in the form of extremely high-energy particles that originate mainly from distant supernovae explosions that occurred millions of years ago. Their energy is great enough to easily penetrate the Earth's protective atmosphere and magnetic field, which would make life impossible by destroying DNA and making our climate inhospitable. The protection provided by the Sun is an interplanetary magnetic field known as the heliosphere. Our Sun is a magnetic star whose plasma atmosphere (mostly electrons, protons, and alpha particles—helium nuclei) blows from its surface to cause solar winds. The solar wind deflects about 90% of galactic cosmic rays away from Earth and our own atmosphere and magnetic shield take care of almost all rest to the point that they present almost zero harm to life.

Plate Tectonics

The map of the world once differed markedly from what we see today. The Earth was once a massive supercontinent that slowly broke into the continents that we now recognize. The mechanism that drove Earth's geological/geographical evolution is known as plate tectonics. Earth's lithosphere, consisting of the crust and upper mantle, is crisscrossed by a jigsaw of large and small rigid plates moving in slow motion relative to adjacent plates. The motion of the plates causes deformation of the land by subduction, a process that occurs when one crustal plate sinks beneath another. This dynamic movement is generated by convection currents, which are flowing fluids that are moving because of temperature or density differences within the material carried from the Earth's interior to its surface. The Earth's heat comes partly from what is left over from its early molten state and partly from the decay of radioactive elements in the core and mantle. Geophysics research has demonstrated that the composition of these radioactive elements is fine-tuned for long-term and stable plate tectonics and the Earth's habitability.

Over millennia, the movement of the continental plates has pushed the Earth's crust up to form magnificent mountain chains and brought needed molecules to the Earth's surface via volcanic activity. Mountains are vital for life because without them there would be no rivers and thus precious little fresh water. Mountains capture and hold snow and ice during the winter and when snow and ice melt in the summer, water cascades down to facilitate our agricultural and industrial activity. We like the look of mountains, but often see them as impediments to travel. We seldom think of the benefits of mountains, but they provide "60–80% of the world's freshwater resources for domestic, agricultural, and industrial consumption… The mountains we see today also supply important minerals and genetic resources for major food crops" (Kohler, Pratt, Debarbieux, Balsiger, et al. 2012, p. 7). We see mountains everywhere on the planet instead of a flat, lifeless water world, so we can thank God for plate tectonics that continuously push them up.

When we look at the majesty of a mountain range, such as the Rockies, we may think that it has always existed, but mountains that existed early in the Earth's history have eroded away long ago. Geologists using a simple formula calculate that a typical mountain mass is 1.2 miles in height and 2.5 square miles wide, and note this mass would have completely disappeared in 123 million years; a very short time in geological terms. But we see mountains everywhere on the planet instead of a flat, lifeless water world. Some gigantic force has to be continuously creating them, and plate tectonics is that force.

Both plate tectonics and the magnetic field depend on convection by which the less dense material of the Earth rises and more dense material sinks. Rocks, water, and air become less dense as their temperature increases, and so they rise; when they become colder, they become denser and sink. It is this convective motion that over time breaks the tough lithosphere into plates. A planet lacking convective activity lacks plate tectonics and all the benefits the process involves. Just as large-mass planets harbor too much gravity for mountains and continents to form, small-mass planets lack plate tectonics, which has the same result. A low mass planet leads to early cooling from its molten state and its crust solidifying into a "stagnant lid" (no movement of its crust). If the planetary mass is too great, plate tectonics may be too mobile. The Earth's plate tectonics have just the right amount of vigor because it "falls within a zone of transition between 'hard' stagnant lid and mobile plate regimes" (Valencia, O'Connell, and Sasselov, 2007, p. 47). Earth thus has a just-right mass.

Without plate tectonics constantly pushing material upwards and maintaining volcanic activity the Earth would be a lifeless water world. We need water, but we don't owe our existence to just Earth's water: "The Earth is not unique because of its oceans. Any planet in the right part of the habitable zone will have those. What is unique about the Earth is that it has LAND. If the moon had not carried away most of the crust, there would be no ocean basins, no land, and no chance for life to evolve on land" (Hoffman, 2001, np). Note that Genesis 1:9-10 described first the appearance of water on the planet, and then the land: "And God said, Let the waters under the heaven be gathered together unto one place, and let the dry land appear: and it was so. And God called the dry land Earth, and the gathering together of the waters called the Seas: and God saw that it was good."

It wasn't until 1963 that science conclusively accepted the "outlandish" idea that the Earth's interior actually moved to "let the dry land appear" and "that it was good." Wallace Pratt, the twentieth century's most prominent geologist, has noted the accuracy of Genesis given its intended audience: "If I as a geologist were called upon to explain briefly our modern ideas of the origin of the earth and the development of life on it to a simple, pastoral people, such as the tribes to whom the Book of Genesis was addressed, I could hardly do better than follow rather closely much of the language of the first chapter of Genesis" (in Copithorne, 1971, p. 14). Pratt further noted that the sequence of events according to the science of geology—first there was water, then the emergence of land, then marine life, followed by land animals, is the same as the ordering revealed in Genesis.

Volcanoes and Photosynthesis

Volcanoes are the result of the movement of the Earth's plates. These fiery cauldrons are destructive, but they also play an important role in Earth's habitability. When volcanoes erupt, lava, and the particulate matter released into the atmosphere, contains many vital minerals, such as phosphorus. Phosphorus is an essential component of life because it supplies part of the sugar-phosphate backbone of DNA and RNA. It is also an essential ingredient of ATP (adenosine triphosphate), an organic compound that provides the energy needed for cellular activity, and is the nutrient that regulates the rate of plant photosynthesis. Volcanic gases are the source of most of Earth's water, most of the atmospheric gases the Earth needs, and provide us with fertile soil.

Because these gases created more than 80% of the planet's surface and laid the foundations for life, Maya Wei-Haas has described volcanoes as Earth's geologic architects: "Their explosive force crafts mountains as well as craters. Lava rivers spread into bleak landscapes. But as time ticks by, the elements break down these volcanic rocks, liberating nutrients from their stony prisons and creating remarkably fertile soils that have allowed civilizations to flourish" (2018, np).

The most beneficial role played by volcanoes is in the formation of the Earth's atmosphere. Early in the planet's history, volcanic gasses stored in its interior collected around the surface and changed the atmosphere from one consisting of hydrogen sulfide, methane, and many times the carbon dioxide that we have in today's atmosphere. Much of the carbon dioxide dissolved into the oceans, where bacteria consumed it and release oxygen as a byproduct. Volcanism regulates the amount of carbon dioxide in the atmosphere by moving carbon in and out of the Earth's interior. Without this process, too much carbon dioxide (a greenhouse gas) would become trapped in the atmosphere, and the Earth would heat up so much that the Earth would become a hot, dead planet. Too little carbon dioxide and the Earth's heat would escape, and it would become a cold, dead, planet. The Earth has many biochemical cycles by which organic and inorganic substances are transformed back and forth for the benefit of life.

Photosynthesis is the process by which plants obtain their food and animals their oxygen. Photosynthesis is vital for life, but it requires demanding constraints that must fall within the right ranges of light intensity, temperature, humidity, carbon dioxide, seasonal stability, water, and minerals. Plants need sunlight to photosynthesize, but most thrive better in softer, shaded, sunlight than in its direct glare. This is where volcanic eruptions come into play. There are at least 20 volcanoes actively erupting at any given moment. Scientists looking at the effects of the 1991 Mount Pinatubo volcanic eruption in the Philippines found that global photosynthesis was increased by this eruption and that it resulted in an average 0.5°F reduction in global temperature. Sulfur dioxide in volcanic ash decreases temperatures by producing sulfur-rich particles called aerosols. Photosynthesis increased because these minute particles in the atmosphere scatter and absorb sunlight. As noted, most plants grow better in diffuse light than in direct light and the volcanic particulates scattered the light globally for two or three years resulting in enhanced photosynthesis (Gu, Baldocchi, Wofsy, et al. 2003).

The Moon

Our Moon is the "lesser light" designed to "rule the night" (Genesis 1:16). It is an anomalously large moon in comparison to its parent planet at almost one-third the size of the Earth. But it is not just this ratio that is odd; a recent in-depth analysis of our Moon and moons found orbiting other planets revealed that a moon such as ours may be extremely rare throughout the known universe (Vieru, 2011). It is generally accepted that the Moon was formed about 4.5 billion years ago by an oblique collision of a Mars-size planet with the embryonic Earth. This "big whack," as it has been dubbed, threw off a massive amount of material that eventually coalesced to the bright body in the night sky. The gravitational pull of the newly formed Moon slowed the Earth's rotation, slowly lengthening our day from about 5 or 6 hours to the current 24 hours. Rotation causes wind, and a rotational period of 5 or 6 hours would have meant constant cyclonic winds raging around the Earth. The big whack collision also helped to create both the magnetic field and plate tectonics. The collision generated such intense heat that liquid iron sank to Earth's center providing the mechanism for generating magnetism, and removed a large amount of Earth's crust, which is important because a much thicker crust may have prevented plate tectonics.

The Earth's spin axis (the degree to which it tilts toward or away from the Sun) varies little from its present angle of 23.5 degrees, but without the Moon's gravity tugging at it, it would vary chaotically. If the tilt was around 60 degrees it would mean that all of the Northern hemisphere would lean toward the scorching heat of the Sun and perpetual summer light for half of the year and the Southern hemisphere in darkness and freezing cold in a six-month winter. Complex life could not exist under those conditions. On the other hand, a tilt of much less than 23.5 degrees would prevent the distribution of wind patterns and hence the distribution of rain around the world. As it is, our stable tilt gives us our predictable four seasons and wind patterns. Physicist Joseph Spradley provides us with some insight into the importance of the Moon for life on Earth and its theistic implications. He notes that the uniqueness of the Earth-Moon system violates the notion that life should be commonplace in the universe. "For Christians, it supports the belief that God can work through natural and seemingly random processes to achieve his purposes in creation. It encourages a new appreciation for the special gift of life and an environment suitable for its survival" (Spradley, 2010, p. 273).

Jupiter: Destroyer and Maker of Planets

Few people are aware of the Moon's vital contribution to life on Earth, and even fewer are aware of the role of Jupiter, the largest planet in the solar system. Jupiter is a gas giant 10 times larger than the Earth and 300 times its mass. A number of astrophysicists have shown that our solar system, for a variety of reasons, is an "oddball" system. The most common mode of planetary formation generates planets with much greater masses than the Earth and are tightly packed closely around their parent star with short orbital periods; both conditions expose planets to deadly radiation and lead to tidal locking. Jupiter was the first planet to form and was able to dominate the planet-building process thereafter. Where the rocky planets are today there were nascent planets destined to be gas giants like Jupiter and Saturn and uninhabitable.

This model of planet formation is known as the "Grand Tack" model, which posits that Jupiter is both a maker and destroyer of planets. Jupiter migrated in, and then back out of the inner solar system early in its history. The gravitational interactions during this long journey caused young planets to crash into each other with the debris either falling into the Sun or leaving the remnants to form the asteroid belt located between Mars and Jupiter. Jupiter itself would have fallen into the Sun if it maintained its inner trajectory, but the formation of Saturn caused a gravitational interaction and fixed their current orbits (Batygin and Laughlin, 2015, p. 4217).

This is science's best explanation of our "oddball" solar system and how catastrophic events caused the debris to coalesce into the small rocky planets, including Earth, and how Jupiter and Saturn (indirectly as the savior of Jupiter) have contributed to the Earth's formation and habitability. Some of the debris (asteroids) of failed planets was flung off into space when it came too close to Jupiter's orbit. Millions of these asteroids and meteorites pummeled the early Earth over a period of about 400 thousand years in what is known as "the period of heavy bombardment," seeding it with vital life-essential elements. We needed an asteroid belt neither too large nor too small. A much smaller belt would have failed to bring the essential chemicals and minerals to us; a larger one would have resulted in Earth being pummeled too much.

Although there are fewer asteroids in our neighborhood today, we need protection from them. The respective masses of Mars and Jupiter, the two planets marking the boundary of the asteroid belt, are most helpful in this regard. If Mars was more massive, as all models of planetary formation say that it should be, its gravity would sling more of these things our way but instead, Mars takes a number of hits for us. But it is Jupiter that serves as our best protector. Jupiter's powerful gravity functions as a giant vacuum cleaner sucking up asteroids that have been bumped out of orbit. Jupiter's clean-up role was in evidence when Comet Shoemaker-Levy 9 impacted it in 1994; had its gravity not sucked it up it may have entered our neighborhood. The solar system's bully has become the Earth's bodyguard. This Jupiter Grand Tack scenario explains critical "unnatural" anthropic features of our solar system for scientists. For Christians, it shows that it is fine-tuned to an exceptional degree as to imply the miraculous hand of God.

Although we need Jupiter, its huge mass, and that of the other gas giants, would have a profoundly destabilizing effect on the resonances of the inner planets in the absence of the Earth and Moon. Kimmo Innanen and his colleagues address the stabilizing role of the Earth-Moon system, stating: "The classical problem of the long-term stability of the solar system continues to motivate, challenge, and perplex dynamicists." Their computer modeling led them to conclude: "Our basic finding is nevertheless an indication of the need for some sort of rudimentary 'design' in the solar system to ensure long-term stability. One possible aspect of such 'design' is that long-term stability may require that terrestrial orbits require a degree of irregularity to 'stir' certain resonances enough so that such resonances cannot persist" (Innanen, Mikkola and Wiegert, 1998, pp. 2055 and 2057). They put "design" in scare-quotes because without them they would have to address who the Designer is.

CHAPTER EIGHT

Avoiding Anthropic Fine-Tuning with the Multiverse

The Multiverse

Physicist Freeman Dyson once remarked: "As we look out into the Universe and identify the many accidents of physics and astronomy that have worked together to our benefit, it almost seems as if the Universe must in some sense have known that we were coming" (Dyson, 1979, p. 250). This is phrased in anthropic terms, terms that some physicists love, some despise, and others ignore. To avoid the fine-tuning with its insane improbabilities some physicists have turned to the extravagant speculations that our universe is but one of an infinite number of other universes in a multiverse. As eminent physicist Paul Davies writes about this war between the Anthropic Principle and a multiverse:

> Scientists have long been aware that the universe seems strangely suited to life, but they mostly chose to ignore it. It was an embarrassment – it looked too much like the work of a Cosmic Designer. Discussion of the anthropic principle was frowned upon as being quasi-religious…. What made a difference was the idea of a multiverse, which offers the opportunity to explain the weird bio-friendliness of the universe as a straightforward selection effect, without invoking divine providence. (2007, p. 151)

The reasoning behind the multiverse conjecture is that because infinity finds us in mathematics, physical infinity must also be realizable. Hypothesize enough universes and you will beat the odds of finding one with its parameters fine-tuned to such an incomprehensible degree as ours. That is, if there are untold trillions of universes then at least one should contain all the "coincidences" that have led to intelligent life on Earth. This is like saying that if you buy all the lottery tickets you are bound to win the lottery. For the multiverse proponent, the Copernican Principle has taken a further giant step: not only is Earth not special or unique, but the universe in which it resides is merely one of countless trillions of others.

The multiverse hypothesis counters the argument from design. If you don't want God, posit a multiverse, because either a fine-tuned universe has a fine-tuner, or else we have a multiverse of trillions of universes in which every possible combination of physical constants and forces exist somewhere. The design notion is that the probability of functional higher-order complexities produced by step-wise interactions of simpler constituent parts is highly improbable when considered against the vastly greater probability that they would produce a huge number of non-functional combinations. The multiverse hypothesis allows the design argument to be rejected because given an infinite multiverse and infinite time, the notion of impossibility disappears; the impossible becomes probable, and the probable becomes inevitable.

Of course, we have no conceivable way of observing other universes and thus cannot prove their existence. Alan Lightman concedes this but says: "Not only *must* we accept that the basic properties of our universe are accidental and incalculable. In addition, we *must* believe in the existence of many other universes... Thus, to explain what we see in the world and in our mental deductions, we *must* believe in what we cannot prove" (2011, pp. 38-40). Lightman sounds like atheism's pope speaking *ex cathedra* and demanding that all devout materialists must believe in an unseen and unknowable multiverse. He, and others like him, have invented this multiverse metaphysical entity that we can never know to get rid of another we can—God.

There are different multiverse models, but those proposed by Max Tegmark are known best. His models are arranged in a hierarchical fashion such that subsequent levels encompass and expand on the lower levels. Level I are universes with the same laws and constants as ours with similar, and even identical, configurations, and thus each of us will have identical twins residing in them. All these parallel universes are said to have arisen when rapid inflation milliseconds after the Big Bang created different universes in bubbles of space with identical laws of physics. We will never be able to see these other universes or hold Zoom meetings with our identical twins since they are beyond our Hubble volume (that portion of the universe that is observable).

Level II models propose universes with different laws of physics, as opposed to universes in Level I where each universe has a different initial distribution of matter but the same laws of physics. Level II assumes that different regions of space exhibit different laws of physics with an infinite number of developmental possibilities. This model posits that the Big Bang was just one of the trillions of space-time bubbles arising within a larger system, and our universe is just one of them. Because our universe allegedly bubbled into existence from a pre-existing mega-universe, it takes us back to a past eternal multiverse. As noted previously, this is not possible according to the second law of thermodynamics, because everything would be in thermodynamic equilibrium by now. But because infinity is mathematically realizable and the multiverse notion is based purely on mathematical models, Tegmark can fiddle with infinity all he likes.

Level III are quantum universes unfolding in every possible way. A quantum system exists in all possible states at the same time (in superposition), and

Tegmark argues that quantum superpositions are not confined to the micro world. He says this because we, and everything else, are made of atoms, and if atoms can be in more than one place at the same time, then so can we (don't let your boss know this)! Tegmark informs us that the only difference between Level I and Level III is where your twins reside. In Level I your twins live in a familiar 3D space like us, but in "Level III they live on another quantum branch in infinite-dimensional Hilbert space" (Tegmark, 2009, p.8). Hilbert space is a mathematical concept used to infer dimensions beyond the familiar four dimensions of everyday reality—three spatial and one-time dimension. Each quantum state is an element of an infinite number of Hilbert spaces. You cannot walk your dog in any space other than 3D space, but Hilbert space is a concept that finds use in quantum mechanics.

In Level IV, universes are made up of any mathematical structure mathematicians can conceive and are governed by different equations from those describing our universe. Level IV is Tegmark's favorite because he argues that any conceivable universe is subsumed within it and there can be no fifth level. He explains that this level "can be viewed as a form of radical Platonism, asserting that the mathematical structures in Plato's realm of ideas…exist 'out there' in a physical sense, casting the so-called modal realism theory…in mathematical terms akin to what Barrow refers to as 'π in the sky'" (2009, p. 12). Modal realism is the view that all possible worlds are just as real as the actual world that we know.

Tegmark thus maintains that mathematical structures have a physical reality outside our familiar spacetime, even though no measurement could ever falsify their existence. Tegmark believes in the Platonic notion that mathematics is the ultimate reality, and that the tangible things they describe are imperfect copies of their real form found only in their mathematical description. He says: "I argue that it means that our universe isn't just described by math, but that it is math in the sense that we're all parts of a giant mathematical object, which in turn is part of a multiverse so huge that it makes the other multiverses debated in recent years seem puny in comparison" (Tegmark, 2014, np). Mathematics is a deductive enterprise with its own notions of truth-seeking separate from empirical science. Multiverse models may appear in math, but testing theories in science rests on induction from experimental data to theory and adjusting the theory as the data warrant.

M-Theory

The mathematical basis for the notion of the multiverse is M-theory, which is a theory that unites the five versions of string theory that scientists have been struggling with for over 50 years. String theories attempt to unify gravity with quantum mechanics by smoothing out the mathematical inconsistencies between quantum theory and the general theory of relativity. M-theory asserts that the fundamental constituents of physical reality are not the particles of standard physics such as quarks, but even tinier filaments of energy called strings. These strings are vibrating energy in a quantum field that gives rise to all particles and forces in the universe. The standard subatomic particles are simply different vibrations of a fundamental string, some of which are useful, such as those in our universe, and others, such as in the vast majority of other universes, are not. They are said not to only vibrate in the three dimensions of space and one of time, but rather in 11 dimensions—10 spatial dimensions plus time—that are "folded" in on one another. These postulated dimensions are curled up in "internal space" in all possible ways, each of which is assumed to be able to describe phenomena with its own restricted range. It has been said that the number of possible solutions to the equations of M-theory may be as many as 10^{500}, which means that any experimental result could be consistent with the theory and it could not be proved either right or wrong.

In their book, *The Grand Design*, Hawking and Mlodinow agree that our universe is exquisitely fine-tuned, but like Tegmark, they want to attribute it to blind luck because if multiple trillions of universes exist, there must be a winner in the ultimate Powerball game. Hawking and Mlodinow inform us that: "People are still trying to decipher the nature of M-theory, but that may not be possible" (Hawking and Mlodinow, 2010, p. 117). However, they continue as though it is not only possible but has been done and dusted. They posit the existence of many different universes by appealing to the "laws" of M-theory (which they previously said may be undecipherable) as existing in internal curled spaces. They write: "The laws of M-theory, therefore, allow for different universes with different apparent laws, depending on how internal space is curled. M-theory has solutions that allow for many different internal spaces, perhaps as many as 10^{500}, which means that it allows for 10^{500} different universes, each with its own laws" (Hawking and Mlodinow, 2010, p. 118).

M-theorists seemingly ascribe intelligence and agency to mathematical equations, since they believe that their equations can bring universes into existence: "Because there is a law like gravity, the universe can and will create itself from nothing... Spontaneous creation is the reason there is something rather than nothing, why the universe exists, and why we exist. It is not necessary to invoke God to light the blue touch paper and set the universe going" (Hawking and Mlodinow, 2010, p. 180). The universe creating itself from nothing means that something that once did not exist managed to create itself. This mysterious nothing is gravity. Gravity is the warping of spacetime, and spacetime warps because of matter, so the concept may be meaningless without matter (although physicists debate this). Because Hawking and Mlodinow insist that gravity existed prior to the existence of the material universe, they should explain how this is possible.

To be fair, Hawking and Mlodinow say that it was the laws of gravity that existed before space-time and matter, not gravity itself. In this view, mathematics has within its nature the power to breathe fire into its own equations and is gifted with its own special brand of agency. Senior NASA astronomer Seth Shostak ridicules the notion of gravity's laws as the prime mover of the universe as Plan B. He notes that Hawking and Mlodinow's assumption that it is all about gravity begs the question of who designed gravity: "Isn't it remarkable that this gentle force seems so perfectly suited to the job of assembling a grand and habitable universe? And indeed, even leaving gravity aside, there are many other physical parameters that seem to be nicely adjusted for our presence. …Depending on your personal philosophies, you can either credit this custom fitting to the intentions of God, or go for Plan B" (Shostak, 2011, np).

By pitting the law of gravity against God who created them and opting for Plan B, John Lennox accuses Hawking and Mlodinow of committing a classical category error of confusing abstract law with personal agency. Laws are mathematical models that describe the behavior of forces or things that exist; they do not possess agency to bring those forces or things into existence. An abstraction has never created a concrete reality. As Lennox said in one of his YouTube debates: "The laws of arithmetic tell me that 2 + 2 = 4, but that has never put £4 in my pocket." A better question would be to ask who made the universe in such a brilliantly fine-tuned intelligible way that it can be described by those elegant mathematical laws. The divine law that Hawking and Mlodinow say created trillions of universes from nothing is gravity, which is something, not nothing. When asked where gravity came from, Hawking answered: "M-theory" (Lennox, 2011, p. 39). So gravity was created by math equations coming to life! Tim Radford, nicely captures the metaphysical God-like nature with which Hawking and Mlodinow endow M-theory: "M-theory invokes something different: a prime mover, a begetter, a creative force that is everywhere and nowhere. This force cannot be identified by instruments or examined by comprehensible mathematical prediction, and yet it contains all possibilities. It incorporates omnipresence, omniscience, and omnipotence, and it's a big mystery. Remind you of Anybody?" (Radford, 2010, np).

M-Theory and Hypotheses in Science

Scientists earn their living seeking new knowledge about the world, but first, they must master what is already known and organize it in a systematic way by fitting facts into coherent and harmonious patterns we call theories. In addition to being looking backward to fit known facts into a coherent pattern, theories must also be forward-looking, telling researchers where they might look to fill gaps in their knowledge. Looking to fill the gaps in our knowledge takes the form of a series of statements that can be logically deduced from a theory called hypotheses, which are deductive statements about relationships between and among factors we expect to find based on the logic of our theories. Theories provided the raw material (the ideas) for generating hypotheses, and hypotheses support or fail to support theories by exposing them to empirical testing. A theory is never proven true, but it must have the quality of being falsifiable (a necessary, but not sufficient requirement for a scientific theory). If a theory is formulated in such a way that no amount of evidence could possibly falsify it, it is not a scientific theory.

Hawking and Mlodinow imply that M-theory possesses predictive accuracy and predictive scope, but it has not provided one scrap of empirical evidence; it is a gun that's never been fired so we cannot gauge its accuracy. Furthermore, there is no way, even in principle, that the theory could be falsified. Failure to discover strings could, but could strings ever be discovered? It is hardly likely because the smaller the hypothesized particle, the more the energy needed to detect it, but even the 17-mile Large Hadron Collider is not sufficient. There is a formula to calculate the radius of curvature needed for a collider to detect the hypothesized strings, as well as for the strength of the magnetic field for the bending magnets that keep particles traveling on a circular trajectory. Physicist Frank Heile (2016) has done the math and shows that the radius would have to be 517 light years, and that the magnetic power required would be quadrillions more powerful than the Earth's magnetic shield. Thus, the multiverse is a highly speculative scenario that is untestable, even in principle. Physicists of an atheistic bent feel that the multiverse is the only way that they can escape God. Many other physicists dismiss the whole multiverse notion, with some claiming that it threatens the scientific status of physics.

There are scientists interested in the mathematics of M-theory who may not give multiverse speculations the time of day because they hope that it may lead them to reconcile relativity theory with quantum mechanics. They realize that M-theory is not a theory in the scientific sense, because such a theory demands empirical support. M-theory is a purely mathematical theory with a ridiculously large number of possible solutions. Early scientists such as Copernicus, Galileo, Kepler, and Newton knew that the universe was capable of mathematical description because a rational God fashioned it that way. However, M-theorists want to decouple mathematics from empirical validation, claiming that the validity of their mathematical models depends on the beauty or elegance of their equations, not on empirical findings. But just because the world is *describable* by mathematics does not mean that the world *is* mathematics. Yet in the rarefied corridors in which M-theory resides, some physicists seem to have reached the conclusion that it is. Their operative equation is evidently that beauty = truth and have quarantined the problem of empirical validation behind a wall of daunting equations.

If physicists invent an imaginary world with clearly expressed beautiful mathematical rules, we can use these rules to gather further mathematical evidence about that imaginary world. From this base, they can shape a theory that is internally consistent according to observers who understand the rules of the mathematics involved and can conclude that because the theory is beautiful it must be valid. As elegant as M-theory is, it is a creature of the imagination designed to explain away the reality of the fine-tuning of our universe. Mathematician George Ellis believes that it is our current inability to explain why the fundamental physical constants have the values they do that drive multiverse speculations. If we could explain them, "the drive for a multiverse explanation would fall away" (Ellis, 2011, p. 295). The math in M-theory is so difficult that it keeps the critical non-mathematician at arm's length. But recall that Herbert Dingle told us that we can tell lies as well as truths with it, and how Einstein stopped the universe in its tracks on paper by inserting lambda into his equations.

Kurt Gödel's version of Anselm's ontological proof of God's existence expressed in the language of mathematical logic is an example of what can be accomplished with mathematics. No one has ever found inconsistencies in Gödel's theorem, and in 2014, computer scientists fed it into high-powered computer programs called "higher-order automated theorem provers" and proved Gödel right (Benzmüller and Paleo, 2014). Of course, multiverse proponents would say that the theorem was proved only according to the internal consistency of the mathematics, which is based on certain assumptions, and they would be right. We cannot prove God exists by mathematics any more than we can prove the existence of a multiverse. Since God is outside of nature, He lies outside of definition and measurement. We can describe millions of things with math because they are material and natural while God is immaterial and supernatural and not amenable to measurement. God is not a *theorem* to be proved; He is a self-evident *axiom* from which everything must be ultimately deduced.

Physicists and Mathematicians Weigh in on M-Theory

M-theory has many supporters, but also many detractors. Nobel laureate physicist Sir Roger Penrose describes Hawking and Mlodinow's multiverse notions as "hardly science," and "not even a theory" (Penrose, 2010, np). Another Nobel laureate physicist, Richard Feynman, dismissed M-theory as "crazy," "nonsense," and "the wrong direction" for physics (in Varghese, P. p. xviii). Mathematical physicist Peter Woit wrote a book-length criticism of string theory, likening it to that caricature of learning called postmodernism. Woit says: "There is a striking analogy between the way superstring theory research is pursued in physics departments and the way postmodern 'theory' has been pursued in humanities departments. In both cases, there are practitioners that revel in the difficulty and obscurity of their research, often being overly impressed with themselves because of this" (Woit, 2006, p. 207).

Ellis and Silk note that the mathematical elegance of M-theory generates grand but untestable hypotheses, and conclude that because M-theory is metaphysical, "theoretical physics risks becoming a no-man's-land between mathematics, physics and philosophy that does not truly meet the requirements of any" (Ellis and Silk, 2014, p. 321). M-theorists, faced with fundamental difficulties in meshing their theories to the observed universe, have argued for a change in how physics is done. Relaxing the criteria by which a theory is judged useful harms physics because among the time-honored criteria for a scientific theory is that it must be falsifiable. Tom Hartsfield joins the chorus of physicists criticizing M-theory and calls it a passing theoretical fad asking that the rules be changed to accommodate it: "To loosen the principles of our fantastically successful scientific method just to allow for one to continue would be a disaster" (Hartsfield, 2016, np).

What if the Multiverse Exists?

But what if against all odds M-theorists turn out to be right and the laws of nature in our universe turn out to be just local by-laws and other localities have different ones? What are the implications for belief in God? Many atheistic scientists affirm that we must make a choice; an infinite number of universes ruled by blind chance or just the one designed by the Creator. When scientists talk of chance outcomes, they are talking about the probability of preexisting forces and preexisting matter influencing an outcome, so there is nothing inherently atheistic at all about attributing outcomes to chance. There are theistic scientists who maintain that it is reasonable to assume that God allows chance to play its part in creation thereby enhancing His respect for the freedom He gives his creation.

The finite mind of man cannot presume to know how an infinite and transcendent God decided to create everything. God can work through random processes to achieve his purpose. To deny that the Lord cannot "work in mysterious ways His wonders to perform" is to question His judgment. Let us not forget the words of Isaiah 55:8-9: "For my thoughts are not your thoughts, neither are your ways my ways, saith the Lord. For as the heavens are higher than the earth, so are my ways higher than your ways, and my thoughts than your thoughts." We delude ourselves if we think we know His thoughts and purpose. As Nobel laureate chemist Richard Smalley remarks: "The purpose of this universe is something that only God knows for sure, but it is increasingly clear to modern science that the universe was exquisitely fine-tuned to enable human life. We are somehow critically involved in His purpose. Our job is to sense that purpose as best we can, love one another, and help Him get that job done" (in Overman, 2008, p. 11).

Nobel laureate physicist Steven Weinberg notes: "If you discovered a really impressive fine-tuning...I think you'd really be left with only two explanations: a benevolent designer or a multiverse" (in Gefter, 2008, p. 48). Richard Dawkins claims that Hawking and Mlodinow's *Grand Design* has fatally wounded God, so there is no choice to be made: the multiverse has won. But the multiverse does not rule out God, nor God the multiverse. To posit trillions of universes to explain the fine-tuning of our universe rather than God is irrational; there is no reason that one should logically preclude the other. M-theorist Gerald Cleaver does not believe that the theory precludes God. He writes: "The bulk universe [the multiverse] is consistent with belief in a God whose nature does not change and whose nature contains the attribute of creating. It yields a picture of an infinite, eternal God, who eternally creates and creates infinitely. This should be no surprise, for those who believe in an eternal, self-consistent God, characterized by all of the classical 'omni' attributes" (2006, p. 7). Bernard Carr also believes that the choice between God and the multiverse is wrong-headed because if God created one universe, he is quite capable of creating many. Yet he finds it: "not surprising that the multiverse proposal has commended itself to atheists. Indeed, Neil Manson has described the multiverse as 'the last resort for the desperate atheist.' For if ours is the only universe, then one has a problem explaining the fine-tunings and might well be forced into a theological direction" (Carr, 2013, p. 168). The multiverse may be "the last resort for the desperate atheist," but even if it exists it provides them no comfort because it still needs a Creator.

Even if a multiverse exists, it is more reasonable that it does so by God's creative hand than by investing God-like power to a mindless something that spontaneously created itself from nothing. Perhaps God purposely created the universe (or multiverse) in such an incredibly unlikely way so that we never stop looking for His fingerprints. If there are trillions of other universes, it only adds to the majesty of God. The things we currently find utterly improbable or even impossible, may in some distant future be found true. And if they are, we can rejoice that they will be found to be the fruits of the design inherent in the laws of the universe that He set in motion. God did not create a universe incapable of being described in natural terms. He is the Agent that designed it all, and who gave us the intelligence and motivation to figure it all out. God's hand is seen in the secondary causes through *His* laws of nature. The "how" questions of nature are the domain of science; the "why" questions are the domain of God's agency and purpose. Confronted with all the evidence we have from science, all we can do is reason to the best explanation of why we are here, and that explanation points unerringly to a creator God of this universe, or of multiple others as well.

CHAPTER NINE

The Molecules of Life

Wonderful, Weird, Water

The Earth teems with anthropic "coincidences," with water being the first we shall look at. Although 60 chemical elements are found in the human body, roughly 96% of its mass is made up of just four: oxygen, hydrogen, carbon, and nitrogen. Most of the oxygen (65%) and hydrogen (10%) are predominantly found in water, which makes up about 60% of body weight. Water and carbon (18% of the body's chemical elements) are synonymous with life. Commenting on the many life-giving anomalies of water and carbon, Harvard biochemist Lawrence Henderson notes: "The chance that this unique ensemble of properties should occur by 'accident' is almost infinitely small. The chance that each of the unit properties of the ensemble, by itself and in cooperation with others, should 'accidentally' contribute a maximum increment is also almost infinitely small" (in Corey, 2001, p. 120). We first look at that marvelous molecule called water.

We take water for granted because it is everywhere, and we do hundreds of things with it. This matrix of life covers almost three-quarters of our planet, and yet no one really understands it. Some scientists spend their careers trying to figure out this marvelous molecule. Water is fascinating because, despite its simple H_2O structure, it is the strangest liquid on the planet because it bends the rules of chemistry. Chemists graph the boiling points of molecules by their atomic weights and find that they all behave as expected—except water. If water "followed the rules" its atomic weight leads chemists to expect, it would boil at -100°C rather than +100°, and at the ambient conditions on the surface of the Earth, it would exist only as a gas.

Atoms and molecules in liquid substances huddle closer together as they get colder and eventually solidify, but water attains its greatest density at just over 39°F, at which point it begins to sink to the bottom of lakes and ponds. Defying all the rules the molecules of other substances obey, water molecules move farther apart as the temperature drops below the level of maximum density and becomes ice, which floats to the top. This insulates the liquid water and the living organisms underneath it. Without this unique property, we would have a runaway freeze-up as layers of ice accumulated from bottom to top, putting an end to aquatic life and all animal life. It would also make the world colder once its mass inevitably reached the surface by reflecting more and more light from the Sun. Water is also unique in that it is the only substance found to exist naturally in solid, liquid, and gaseous forms.

Water is vital for metabolism, temperature regulation, and the flushing of toxins. It is the solvent required for biochemical reactions, affecting the stability, flexibility, structure, and dynamics of proteins, nucleic acids, and DNA. Life depends on many of its anomalous properties, and if it behaved as expected we wouldn't be here. Water is a "polar" molecule that has a tiny positive charge at the hydrogen pole and a tiny negative charge at the oxygen pole. The molecule has a tetrahedral geometry with an atom located at the center with four substituents located at the corners of a tetrahedron (having four plane triangular faces like a triangular pyramid). As a gas, H_2O is lighter than almost any other gas, as a liquid, it is much denser, and as a solid, it is much lighter than chemists are led to expect from the nature of other molecules.

Water molecules are so good at bonding to each other that they can actually overcome the force of gravity by capillary action. Capillary action occurs because water molecules like to stay close together, but they are also attracted to and stick to other substances. This allows water pooled at the root of a plant to move upwards to feed its leaves when the adhesion to the plant walls is stronger than the cohesive forces between the water molecules. This is also the marvelous force that takes water-conveying oxygen and other nutrients up to your brain.

The same water has been recycled on the earth for billions of years moving nutrients, pathogens, and sediment in and out of aquatic environments in the hydrologic cycle. This cycle is a perfect method of distributing freshwater around the planet to plants, animals, and us. We begin the cycle with the heat of the Sun lifting surface water via evaporation as gas. These gas molecules are too small and weak to bring with them major contaminants muddying the waters. However, there is no "pure" water found in nature; H_2O molecules are very sticky and attract all kinds of harmless particulates. Clouds form from these molecules as condensation, and when the air cools to the point that it cannot support more water vapor, droplets of rain, hail, sleet, or snow become precipitation. When these droplets fall to the ground, the water supply is replenished, which is then ready to evaporate again to keep the cycle going via endlessly different paths that water can take on its pilgrimage of life.

Water is crucial for maintaining our planetary atmosphere by both heating and cooling it. The Sun bombards the Earth with a tremendous amount of heat, the majority of which is absorbed by the ocean acting like a massive heat-retaining solar panel. Water has a very high heating capacity, which means that it takes more energy to raise its temperature than it does for other liquids, a quality that allows humans to regulate their body temperature. Water can soak up a lot of heat energy before it changes from liquid to gas, which enables the oceans to store heat during the day keeping us from getting too hot and releasing it at night preventing us from getting too cold. The Sun's radiation is unevenly distributed on the surface of the Earth. The equatorial areas get the lion's share of this heat, but the oceans have a way to distribute this excess heat to colder areas via currents caused by the Earth's rotation, winds, and tides. Like a gigantic conveyer belt, it moves heated water from the hottest areas of the planet to the coldest areas and in return, the hotter areas get cold water from the coldest. Without this movement and temperature regulation, the Earth would be super-hot at the equator and super-frigid toward the poles, which would reduce the habitable landmass on Earth.

An atmosphere at the right temperature supplying the right gravitational pressure on a planet's surface is needed to keep water from boiling away into space. Through its role in temperature regulation, plate tectonics has sustained the conditions required for surface liquid water to exist over billions of years. But since the Earth began as an inhospitable ball of white-hot molten rock, where did all our water come from? Scientists admit that water's origin is still something of a mystery. Some say that planetesimals (asteroids and icy comets) from the outer solar system provided Earth with our water, while others maintain that less than 50% of it could have come from planetesimals and that most of it has been here from the beginning. This notion is that when Earth was forming, hydrogen from the solar nebula (gases and dust that formed the Sun and planets) was incorporated into its interior. Volcanic eruptions and chemical reactions in the Earth's mantle combined hydrogen with oxygen—the most abundant element in the Earth's crust—helped to create Earth's abundant water supply. "As long as the supply of hydrogen can be sustained, one can speculate that water formed from this process could be a contributor to the origin of water during Earth's early accretion. Water formed in the mantle can reach the surface via multiple ways, for example, carried by magma in the form of volcanic activities" (Coghlan, 2017).

Water is everywhere in the cosmos; after all, hydrogen is the most abundant element in the universe, and oxygen is the third most abundant. The abundance of water does not mean that Earth-like planets are abundant. Geologists Jan Zalasiewicz and Mark Williams (2014) marvel at the uniqueness of our planet: "The more we learn about how Earth acquired and retained its water, the more it seems the situation was incredibly fortuitous...Even in a water-filled cosmos, Earth might still be one of a kind amid water worlds far weirder—and more hostile to life—than our own. We might be in possession of an extraordinarily precious, rare jewel: our oceans."

The fine-tuning of water for life is evident at the quantum level. Lisa Grossman states: "Water's life-giving properties exist on a knife-edge. It turns out that life as we know it relies on a fortuitous, but incredibly delicate, balance of quantum forces." Water is held together by weak hydrogen bonds whose lengths keep changing. This should destabilize the network structure of water and remove many of water's life-sustaining properties were it not for a second quantum effect that cancels the effect of the first. One quantum effect involves hydrogen atoms tugging at oxygen atoms in neighboring molecules, pulling them closer, the other is energetically repelling them and increasing the space between the molecules. This latter effect decreases density and makes ice formation possible. Overall, quantum effects keep the H_2O network intact. Grossman concludes by noting: "We are used to the idea that the cosmos's physical constants are fine-tuned for life. Now it seems water's quantum forces can be added to this 'just right' list" (2011, p. 14).

Carbon: The Indispensable, Improbable, Scaffolding of Life

Carl Sagan once famously remarked that we humans are literally "made of star stuff." All the elements in our bodies were forged in the stars, but all carbon is paramount because all life is carbon-based. Carbon atoms form the backbone of millions of organic compounds, the major groups of which are carbohydrates, lipids, proteins, and nucleic acids. Carbon has 6 nucleons (6 protons and 6 neutrons) in its nucleus and 6 electrons swirling around it. The carbon atom has two electron shells, the inner one holding two electrons and the outer one holding four valence electrons. Valence electrons are electrons that determine the number of other atoms with which an element can form covalent bonds (sharing of electrons between atoms) with others. Carbon has room to form 4 bonds with other elements, which will fill its outermost shell with the 8 electrons needed for stability. This configuration allows carbon to form an unparalleled number of bonds and is one of the few elements that can bond with itself. Furthermore, carbon bonds can be formed and broken with a small amount of energy, which easily facilitates the dynamic organic chemistry that occurs in our cells.

All elements in the same column in the periodic table possess four valence electrons and can form a variety of bonds, but none can do so with the ease and stability of carbon. Silicon, the element right below carbon in the table, forms a large number of molecules, but unlike double-bonded carbon, double-bonded silicon is unstable and quickly becomes two single-bonded silicon atoms again. Unlike silicon, the carbon in our bodies contributes crucially to plant life by interacting with oxygen to produce carbon dioxide (a gas formed by a carbon atom covalently double-bonded to two oxygen atoms). Humans exhale carbon dioxide after inhaled oxygen reacts with carbon during respiration. Plant life thrives on carbon dioxide and plants provide us with oxygen in return. When silicon reacts with oxygen it produces quartz, which is of no use to animals or plants. In short, no better element exists for the chemistry of life than carbon.

Like all elements, carbon is produced by the stars, and we have it in abundance (the fourth most abundant element in the universe). Physicists used to be puzzled by why there is such an abundance when most stars are thousands of times cooler than required to burn helium into carbon. Then Fred Hoyle came up with his anthropic prediction of the energy level needed for the stellar nucleosynthesis of carbon. Carbon is made when three "alpha" (α) particles (the nuclei of helium) fuse their combined 12 nucleons (6 protons, 6 neutrons) to form carbon-12 (^{12}C). But "as soon as ^{12}C is synthesized from helium, it absorbs another α particle and becomes ^{16}O [oxygen] leaving no carbon. The reaction forming ^{12}C was much slower than the reaction that destroys it. If so, argued Hoyle, life should not exist!" (Shaviv, 2015, p. 311). John Gribbin and Martin Rees note that Hoyle reasoned from the fact that we exist to predict that carbon must have an energy level at 7.6 MeV (7.6 million electrovolts—the units of energy that provide particle acceleration), and that was precisely the level experiments found it to be: "There is no better evidence to support the argument that the Universe has been designed for our benefit–tailor-made for man" (Gribbin and Rees, 1989, p. 247).

The energy level of 7.6 MeV is Hoyle's prediction of the resonance state of ^{12}C. Resonance refers to the reaction by the excitation of an object's internal motion ("vibration") by an outside source. Every object in the universe, including atomic nuclei, has its own vibration frequency. For instance, if we have two identical tuning forks mounted on sound boxes and strike the first with some object, it begins vibrating at its natural frequency. These vibrations set the air inside the second sound box vibrating at the same frequency because of the sound waves impinging on it. If the prongs of the first fork are grabbed to prevent further vibration, the same sound is heard from the unstruck fork. The incoming sound waves made by the first fork synchronize with the second fork, which begins to vibrate at this shared natural frequency. Another example is the use of ultrasound to break up kidney stones and gallstones. The ultrasound frequency matches the frequency of the stones and causes them to oscillate and break up. This matching of energies is resonance. We use the language of resonance in everyday language; "I feel his vibes;" "That song really resonates with me;" "Grace and I are on the same wavelength."

To get two atomic nuclei to fuse requires the energy of incoming nuclei to resonate with the energy of the receiving nuclei. The transport of energy from one atom to another is called resonance energy transfer. Two fused helium nuclei create a highly unstable isotope of beryllium (^{8}Be) which decays back into two helium nuclei in about 10^{-17} seconds (one ten-thousandth of a trillionth of a second). Making carbon thus requires exquisite precision in timing, and is known as the "triple alpha process." The process involves three steps as shown in Figure 8.1. First, two alpha particles (^{4}He) fuse to form beryllium (^{8}Be) and emit a gamma ray. The ^{8}Be will decay back to helium within 10^{-17} seconds unless a third alpha particle fuses with ^{8}Be to produce the excited resonance state of ^{12}C. As we have seen, when particles come together a small portion of their mass is converted to energy as the strong force pushes them together to overcome the electromagnetic force holding them apart. These are the gamma photons emitted by the ^{4}He/ ^{8}Be fusion. The electromagnetic and strong nuclear forces are invariant physical constants that govern nuclear fusion and have to be perfectly calibrated, both individually and with respect to one another, to make fusion possible. It is estimated only one in 2,500 of these fusions transition to stable carbon atoms; the rest decay. Even the one-in-2,500 ^{12}C atom is in danger if another alpha particle fuses with it to produce oxygen (^{16}O). We need oxygen, but we don't want it at the expense of carbon. Preventing that is their respective energy levels. A very slight change in the nuclear resonance levels of oxygen (MeV 7.12) and carbon (MeV 7.65) would make the production of either impossible.

Figure 8.1. The Triple-Alpha Process

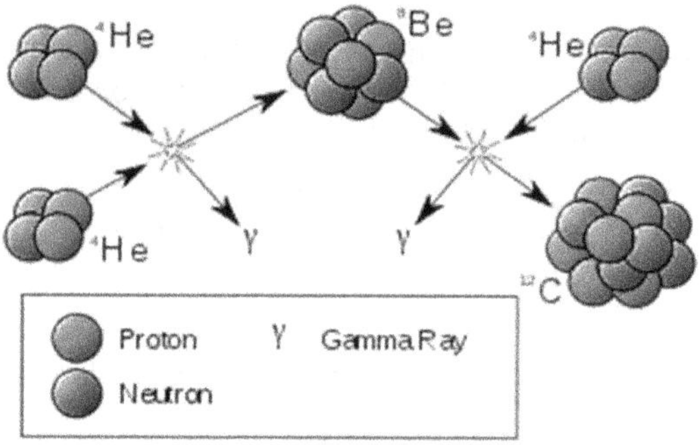

If the Hoyle energy state was a little different there may be an abundance of carbon, but helium would burn into carbon much earlier and the star would not be hot enough to produce sufficient oxygen. Cassé explains: "It turns out that the sum of the mass energies of carbon and helium is just 1% above an energy level of oxygen-16. But this 1% difference is not enough for all the carbon to disappear in the stellar crucible, thereby destroying any chance of life at a later date" (2003, p. 143). George Greenstein marvels at the process of making carbon: "Other nuclear reactions do not proceed by such a remarkable chain of lucky breaks. ... It is like discovering deep and complex resonances between a car, a bicycle, and a truck. Why should such disparate structures mesh together so perfectly? Upon this our existence, and that of every life form in the universe, depends" (Greenstein, 1988, pp 43-44).

Fred Hoyle writes of his awe of the miraculous relation of the energy levels of carbon and oxygen in the *Annual Review of Astronomy and Astrophysics:* "If you wanted to produce carbon and oxygen in roughly equal quantities by stellar nucleosynthesis, these are the two levels you would have to fix, and your fixing would have to be just where these levels are actually found to be." He does not speak of lucky breaks; rather, he asks if it could be "another put-up job," and concludes that: "A commonsense interpretation of the facts suggests that a super-intellect has monkeyed with physics, as well as with chemistry and biology, and that there are no blind forces worth speaking about in nature. The numbers one calculates from the facts seem to me so overwhelming as to put this conclusion almost beyond question" (Hoyle, 1982, p. 16).

Only God could be the super-intellect that "monkeyed" with the laws of physics (carbon's highly improbable fusion), chemistry (carbon's amazing bonding features), and biology (carbon's basis for life)? And which of the interacting "coincidences" out of the thousands found so far will be the final straw to lead the atheist scientist to God? Hoyle himself believed that any scientist who examined the evidence would "draw the inference that the laws of nuclear physics have been deliberately designed with regard to the consequences they produce inside the stars. If this is so, then my apparently random quirks have become part of a deep-laid scheme. If not, then we are back again to a monstrous sequence of accidents" (in Holder, 2013, p. 48). Do you prefer to locate your belief in a "monstrous sequence of accidents" or in God?

Photosynthesis and the Carbon-Oxygen Cycle

Biochemists tell us that: "Photosynthesis is the largest-scale, best-tested method for solar energy harvesting on the planet, and it is responsible not only for the energy stored in coal, petroleum, and natural gas but also for the energy that powers most of the biological world, and for the earth's oxygenated atmosphere" (Gust, Moore, and Moore, 2009, p. 1891). Plants are the first link in the food chain; we eat plants and the animals that eat plants. Plants are living things that grow and reproduce, so they also need nourishment. Unlike animals, plants make their own food internally from non-living materials in the process of photosynthesis. It is impossible to overestimate the importance of photosynthesis for life. It may be the most important chemical reaction on the planet because it not only makes food for animals and plant life; it provides us with the oxygen we need and traps the carbon dioxide we don't. So, if the process stopped because we would literally be "out of breath."

The ingredients plants use to make their food are atmospheric carbon dioxide, sunlight, and water. Carbon dioxide and sunlight enter plants through microscopic holes on the underside of the leaf called stomata, which have antennae that capture the needed carbon dioxide and photons. Water enters the plant through its roots and climbs up to the leaves where photosynthesis takes place. Within each cell on the plant's leaves are organelles called chloroplasts, which both store the energy of the photons and provide plants with a chemical called chlorophyll. Chlorophyll transfers the sun's energy to a photon reaction center in the plant's cells where it is converted to the chemical energy used to split water molecules into hydrogen and oxygen. After this split, the hydrogen is combined with carbon dioxide and used by the plant to produce its starch, sugar, and sucrose food, and the oxygen is released into the atmosphere through the stomata as a waste product. To prevent the energy from dissipating as heat, the transformation of photons into carbohydrates takes place in a miraculous one million billionths of a second.

When a plant's antennae capture photons and transfer the excitation energy to reaction centers, there must be mechanisms to regulate the rate of delivery of the energy long enough to resist natural recombination and for them to perform the work nature requires. Photosynthesis needs such a resisting mechanism because photons that are energetic enough to break H_2O apart would break apart most other biological molecules. These mechanisms rely on quantum mechanics as a way to transport energy in the most efficient way possible. Because of superposition, when plants are in an environment that varies in warmth and moisture, a quantum particle/wave is everywhere at once and thus can take every possible path to its destination in photosynthesis.

Photosynthesis removes large quantities of carbon dioxide from the atmosphere. Without Earth's plant life continually sucking up carbon dioxide, it would build up in the atmosphere trapping heat and causing disastrous climate change. Human industrial and personal activities, such as burning fossil fuels while at the same time destroying large swaths of forest vegetation, are destroying the delicate ecological balance. Photosynthesis, as well as the ultimate source of our food, is responsible for all the energy burned in the world stored in fossil fuels such as coal and petroleum. It thus makes sense to try to harness this energy directly by artificially mimicking photosynthesis, but chemists, biologists, and physicists have been at work since photosynthesis was discovered over 100 years ago trying to duplicate God's ingenuity without much success. They keep trying though because artificial photosynthesis is the holy grail of energy.

Because plants are living things, they respire; the inhalation of oxygen and exhalation of carbon dioxide. Respiration is a metabolic process by which an organism acquires energy through oxidizing nutrients. Although plants respire, they do not have specialized organs like lungs for breathing, nor do they have a circulatory system to transport the gases to their cells. However, they require oxygen for respiration and need to expel carbon dioxide produced during the process. Plants use stomata and lenticels (found in stems) for this gas exchange. Plant respiration is the opposite of photosynthesis in the same way as breathing out is the opposite of breathing in. During respiration, plants use oxygen and the sugars produced during photosynthesis to produce energy for growth. Plants can only photosynthesize when in the sunshine, but respiration occurs constantly. Photosynthesis and respiration occur simultaneously during the sunlight hours but the amount of oxygen produced by plants greatly exceeds that of carbon dioxide.

Leaves are green during the spring and summer due to their high concentration of chlorophyll. Hidden within the leaves at lower concentrations are other pigments that emerge in the fall when temperatures dip and there is less sunlight. This is a stressful time for plants, but they have mechanisms that help them to cope in the form of antioxidant chemicals called anthocyanins (ACNs). These natural colorant chemicals function to maintain a delicate balance between low temperatures and energy management. A plant's life-maintaining enzyme activity occurs more slowly in low temperatures, and when they receive too much light during this time their cells create high-energy forms of oxygen called free radicals. Free radicals are detrimental to all life forms, including humans. This is why consuming foods high in ACNs, such as the berry fruit family, is advised. ACNs divert and store excess photon radiation from excited electron transport chains, thus providing cold-weather resistance for ACN-producing leaves. This finely tuned process keeps plants alive from season to season even as their leaves wither on the ground providing nutrients for the soil (Nassour, Ayash, and Al-Tameemi, 2020). How great is God!

Biologists take it for granted that the process of photosynthesis evolved piecemeal in tiny steps, although there is no direct evidence to support any of the hypotheses that have been advanced for it. Surely, all the functioning parts and the physics and biochemistry involved in the many steps of photosynthesis and respiration must be present simultaneously if it is to occur. Just the assembly of chlorophyll takes 17 enzymes, which require many base pairs in the genetic code to line up correctly. How likely is all this to have occurred by blind evolutionary chance? "Why would evolution produce a series of enzymes that only generate useless intermediates until all of the enzymes needed for the end product have evolved?" asks plant geneticist Rick Swindell. He answers that: "if groups [of bases] of 1,000 recombined at a rate of a billion per second (10^9 tries) for 30 billion years (10^{18} seconds), with the number of bases being equal to the number of electrons that could fit with no space between them into a universe of 5-billion-year radius (10^{130}). This would yield 10^{157} total tries, an inconceivably huge number..." (Swindell, 2003, p. 79).

The Nitrogen Cycle

Nitrogen is the most abundant element in the Earth's atmosphere and is nature's chief fertilizer. Approximately 78% of the atmosphere is nitrogen, 21% is oxygen; the remaining one percent is made up of other gases. Nitrogen is a core component of many plant structures and is often referred to as the "backbone" of plants. It provides energy for their metabolic processes and is an essential element, providing the amino acids for building plant (and human) proteins, and nucleic acids to form DNA. Because of its unique structure, no other element could possibly be substituted to make these vital molecules. Although nitrogen is abundant, it is not accessible to living things directly. The only way we get nitrogen is from the plants we eat, or from the animals that have eaten the plants. The plants themselves obtain it from nitrates, which is made by a series of complex chemical reaction from atmospheric nitrogen.

How nitrogen gets cycled to produce nitrates is probably the most complex of all Earth's cycling processes. Nitrogen gas is very unreactive because it has appropriate proton-neutron ratios, so they don't have to give off energy to stabilize themselves. A tremendous amount of energy is thus needed to turn it into a usable form so that it can be incorporated into our cells in reactive form. We cannot absorb nitrogen directly because its molecules consist of two very tightly bonded atoms that do not readily interact, and the human body's chemistry does not provide sufficient energy to break them apart. The conversion of nitrogen gas into usable form is done for us in two different ways: biological and physical. Biologically, a process called nitrogen fixation is needed to change atmospheric nitrogen (N_2) into usable reactive form for plants. Most nitrogen fixation is done by soil bacteria that possess an enzyme that combines N_2 with hydrogen (H_2) to make ammonia (NH_3) and then nitrates (NO_3).

Another way nitrogen's tight bonds are split is through the awesome power of lightning. A typical bolt of lightning has enough electrical energy (a lightning bolt is about 50,000°F to separate the nitrogen atoms floating in the atmosphere. When they are separated, some of the free atoms combine with oxygen to form nitrogen dioxide. Nitrogen dioxide dissolves in water, which creates nitric acid, which creates nitrates. Nitrates are powerful natural fertilizers that mix with the rain and fall to earth, watering and fertilizing the soil at the same time. No wonder farmers love lightning storms and their nitrogen-charged raindrops! When plants and animals die, their nitrogen compounds are broken down by bacteria as the organic matter decays, which transforms nitrates back to nitrogen gas, which is then released from the soil and back into the atmosphere, thus completing the cycle (Fowler et al., 2013).

CHAPTER TEN

The Queen of all Scientific Problems: The Origin of Life

Life from the Lifeless

Materialist science proposes that life arose from dead matter, and uses the term *abiogenesis* for the hypothetical process by which chemical evolution became biological evolution. In this view, the leap from non-living matter to living matter would require a set of random non-living molecules to arrange themselves, completely undirected, in specific and complex ways to gain the capacity for both metabolism and reproduction. These are the systems that define life, but before life existed, how did these things produced only by living systems, come into being? The challenge for the origin of life (OoL) researchers is not only how the dead become the living, but also which of these systems came first.

Chemistry outside the cell is hostile to it, so living cells must have been enclosed in a protective bubble. They must also have a way of drawing energy from the environment to fuel their many functions and to replicate themselves. These things are so interdependent and complex that it is difficult to imagine how these things arrived in stepwise evolutionary fashion. Francis Crick, the co-discoverer of the structure of DNA has stated: "An honest man, armed with all the knowledge available to us now, could only state that in some sense, the origin of life appears at the moment to be almost a miracle, so many are the conditions which would have had to have been satisfied to get it going" (in Lim, 2017, p. 58). There are many chicken-or-egg problems in OoL research, which explains why 150 theories of abiogenesis were published between 1957 and 2000, and they keep coming (Świeżyński, 2016).

Materialism appeals to immense periods of time to explain the deep chasm between the lifeless chemistry on a dead Earth and the stunning complexity of DNA. This amounts to nothing more than saying that life simply happened, period. Biologists were optimistic about achieving a real explanation when Dean Kenyon, one of the leading biophysicists of the twentieth century, published a book in 1969 claiming that abiogenesis was not only possible but inevitable. This delighted materialists because it ruled out a Creator. However, after 30 years trying to determine how lifeless chemistry could self-organize to become life, Kenyon came to the conclusion: "We have not the slightest chance of a chemical evolutionary origin for even the simplest of cells…so, the concept of the intelligent design of life was immensely attractive to me and made a great deal of sense, as it very closely matched the multiple discoveries of molecular biology" (2002, p. 35). This was an epiphany for Kenyon, who abandoned his atheism for Christianity.

Nobel laureate biochemist Christian de Duve agrees with Kenyon: "If you equate the probability of the birth of a bacterial cell to that of the chance assembly of its component atoms, even eternity will not suffice to produce one for you" (in Andrews, 2017, p. 248). He also wrote:

> Monod stressed the improbability of life and mind and the preponderant role of chance in their emergence, hence the lack of design in the universe, hence its absurdity and pointlessness. My reading of the same facts is different. It gives chance the same role, but acting within such a stringent set of constraints as to produce life and mind obligatorily. ...To Monod's famous sentence "The universe was not pregnant with life, nor the biosphere with man," I reply: "You are wrong. They were." (De Duve, 1995, p. 300)

Another Nobel laureate, George Wald, noted that life from non-life is either chance or Divine creation. He ruled out Divine creation because: "We cannot accept that on philosophical grounds; therefore, we choose to believe the impossible: that life arose spontaneously by chance!" (Wald, 1954, p. 48). Wald later became a deist upon contemplating the fitness of the universe for life and consciousness (mind as an immaterial phenomenon). He wrote that: "This is with the assumption that mind, rather than emerging as a late outgrowth in the evolution of life, has existed always, as the matrix, the source and condition of physical reality—that the stuff of which physical reality is composed is mind-stuff" (Wald, 1984, p. 1). Christians call Wald's mind-stuff "the Word of God."

Amino Acids, Chirality, and Reaction Rates

Early speculations about the OoL posited that life could not have formed in an oxygen-rich atmosphere because oxygen interferes with reactions that transform simpler organic molecules into more complex ones. It was therefore posited that Earth's early atmosphere must have been "reducing;" that is, one in which there is little or no oxygen present and one that easily produces chemical reactions. It was thought to be rich in hydrogen and other compounds such as methane, ammonia, hydrogen, and water that readily donate atoms to other substances. These reducing gases were considered the major components in the "primordial soup" by which chance and necessity produced the basic units of life in the form of the amino acids that build proteins.

In 1953, a famous experiment known as the Miller-Urey experiment was conducted by Harold Urey and Stanley Miller. They created a system of flasks containing the reducing gases assumed to constitute the Earth's early atmosphere. A Bunsen burner served as a heat source, and electrodes provided a continuous electric spark in the apparatus to mimic the role of lightning in the real world. After about a week, a tar-like sludge was produced in the flask which contained five amino acids, the building blocks of proteins, and surmised that that "life-in-the-lab" was just around the corner. But the distance from amino acids to proteins can be measured in light years. Amino acids do not live, and the fact that they combine to make proteins in a very specific way presents a huge problem for OoL researchers.

The problem is that in a reducing oxygen-free world there would be no ozone layer (made by oxygen), and large amounts of ultraviolet radiation would reach the Earth's surface, which would make delicate chemical reactions extremely difficult. Oxygen thus presents a paradox because either its presence (interfering with chemical reactions) or its absence (no ozone protection) stymies prebiotic molecule formation. Furthermore, we now know that the conditions on early Earth were not conducive to the formation of a reducing atmosphere, but rather they were conducive to an oxygen-rich atmosphere such as our current atmosphere. As atmospheric scientist Bruce Watson put it: "We can now say with some certainty that many scientists studying the origins of life on Earth simply picked the wrong atmosphere" (2011, np).

Lab-made amino acids cannot be made to self-assemble into chains to form proteins because proteins are the most structurally complex and sophisticated molecules known to science. Amino acids are monomers ("one part") that must bond together into large molecular chains called polymers ("many parts") to form functioning proteins in the process called polymerization. An amino chain forming in the hypothesized prebiotic soup would have been more likely to break apart than to assemble further. Chemists point out that there is no evidence that a primordial soup ever existed, but even if it did: "Polymerisation into RNA requires both energy and high concentrations of ribonucleotides [the building blocks of RNA]. There is no obvious source of energy in a primordial soup. Ionizing UV radiation inherently destroys as much as it creates" (Lane, Allen, and Martin, 2010, p. 272).

Living things need to extract energy from the environment to continue living, and unguided polymerization runs afoul of the second law of thermodynamics. Because polymerized molecules have already reacted, they are at thermodynamic equilibrium. No further reactions can occur in a system in a state of thermodynamic equilibrium because there is no free energy intrinsic to the system that would allow them to. Free energy can only be supplied to a living system by a mechanism (metabolism) that can harvest energy from the environment to counteract the decaying effects of the second law; only then can the living system break free from its shackles. The problem is that a system must already be alive for it to possess such a mechanism.

Getting amino acids to polymerize and produce a functional protein runs into the so-called chirality problem. *Chiral* comes from the Greek for "hand." Two amino acids that are alike in structure and function may also be distinct from each other because they are mirror images, just as your hands are. One amino acid version is labeled D ("dextro") for right-handed, and the other L ("levo"), or left-handed. D and L amino acids are structurally identical (they have the same atoms: carbon, hydrogen, oxygen, and nitrogen) just as your hands are identical, but you cannot fit your right hand into your left glove. Likewise, D and L amino acids will not bond because chemical reactions that drive our cells only work with molecules of the correct handedness.

Amino acids found in nature come equally in D and L forms. An equal number of D and L amino acids form a *racemic*, but a homochiral set of acids is necessary for life; that is, all amino acids must be left-handed, just as all sugars (ribose) must be right-handed if DNA and RNA are to be produced. Even one right-handed amino acid would destabilize the DNA double helix and it would not be able to form chains of information. Given that the laws of nature always produce a racemic, what is the probability that even a short protein could arise from only left-handed monomers in a racemic on a prebiotic Earth? Plaxco and Gross use a short chain of 189 acids and inform us that it "is highly improbable that random chemistry could produce a polymer molecule that contained monomers of only one-handedness. To be precise, the probability of achieving homochirality in a 189-unit polymer from an equal-molar mixture of left- and right-handed monomers is 1 in 2^{189} (1in 8 x 10^{56})!" (Plaxco and Gross, 2006, p. 114). A 189-unit polymer is very short; many are thousands of monomers long.

There is another problem besides the mind-numbing statistical improbability of achieving homochirality, and that is back again to the second law. Biochemist A. Garay notes: "Consider one of the simplest steps in the origin and evolution of life, the choice of one chiral form over racemic mixture. Thermodynamics do not permit this initial step. They dictate full racemization of all non-completely racemic mixtures. In this respect, weak nuclear interactions seemingly do not obey the second law" (Garay, 1993, p. 168). There is also the problem of reaction rates. If you have different L-amino acids in the lab and allow them to interact, the most reactive acid will link up first and the least reactive will line up last. In other words, the molecules of life are not ordered in any way by amino acid reaction rates. Given the hundreds or thousands of L-amino acids have to line up in a precise sequence to get a functional protein, it is no surprise that this never happens by unguided processes. If only the laws of physics and chemistry determined the sequence we would not be around since getting the precise sequence in an unguided context is beyond unlikely. As molecular chemist

Steven Benner informs us: "An enormous amount of empirical data has established, as a rule, that organic systems, given energy and left to themselves, devolve to give useless complex mixtures" (2014, p. 341). Unguided organic reactions in a pool of chemicals form a gooey tar, a problem known as the "asphalt problem." Benner lists a number of other seemingly unresolvable paradoxes that "suggest that it is impossible for any non-living chemical system to escape devolution to enter the Darwinian world of the 'living'" (Benner, 2014, p. 342).

The Multiverse and Panspermia

Evolutionary biologist Eugene Koonin has calculated the enormous improbability for the simultaneous emergence of translation (ribosomes using RNA as a template to make proteins) and replication (DNA making a replica of itself): "the probability that a coupled translation-replication emerges by chance in a single O-region [observable region of the universe] is P< 10^{-1018}. Obviously, this version of the breakthrough stage can be considered only in the context of a universe with an infinite (or, at the very least, extremely vast) number of O-regions" (Koonin, 2007, p. 19). Koonin's calculation is just the probability of getting replication and translation. You still have to get these functions enclosed in a cell with all its complex interdependent parts. But never mind the messy chemistry; just concentrate on trillions of "O-regions" and blind chance, and problem solved!

Fred Hoyle and Chandra Wickramasinghe wrote of probabilities of getting the 20 amino acids to line up correctly and of obtaining a suitable sugar backbone for DNA/RNA, and the probability of getting functioning enzymes. They concluded: "there are about two thousand enzymes, and the chance of obtaining them all in a random trial is only one part in $(10^{20})^{2000} = 10^{40,000}$, an outrageously small probability ... this simple calculation wipes the idea entirely out of court" (Hoyle and Wickramasinghe, 1981, pp. 19-21). Hoyle looked to panspermia as a way out of the conundrum. There are two versions of panspermia–directed and undirected. Undirected panspermia posits that life arose somewhere in the vastness of the cosmos and hitched a ride on the millions of comets, meteors, and asteroids that bombarded the early Earth. The hostile environment of interstellar space led Francis Crick and Leslie Orgel to dismiss it for directed panspermia, writing: "It now seems unlikely that extraterrestrial living organisms could have reached the earth either as spores driven by the radiation pressure from another star or as living organisms embedded in a meteorite. As an alternative. . . we have considered Directed Panspermia, the theory that organisms were deliberately transmitted to the earth by intelligent beings on another planet" (Crick and Orgel, 1973, p. 341).

The notion of panspermia does not solve the OoL puzzle. It merely moves its origin elsewhere in the vastness of space where the same $10^{40,000}$ problem exists. Even if we conceive of a multiverse with trillions of universes with perhaps 1 in a billion having some potential for life, $10^{40,000}$ still wipes chance "entirely out of court." Hoyle was aware that he had merely moved the origins of life elsewhere, and that does not solve the problem of how life arose, or why it did, but posited "intelligent alien control" over the process. In Hoyle's *The Intelligent Universe*, he wrote: "Even after widening the stage for the origin of life from our tiny Earth to the Universe at large, we must still return to the same problem that opened this book—the vast unlikelihood that life, even on a cosmic scale, arose from non-living matter. It is apparent that the origin of life is overwhelmingly a matter of arrangement by intelligent control. *Unintelligent natural selection is only too likely to produce an unintelligent result* [my emphasis]" (in Korthof, 2006, np). Hoyle left unanswered the nature of this intelligent controller, yet the enigmatic nature of this brilliant scientist made many statements in his books and articles in which we may envision him struggling not to mention God while using metaphors that strongly suggest that he had God in mind. His colleague, Chandra Wickramasinghe, was more forthcoming about God: "From my earliest training as a scientist, I was very strongly brainwashed to believe science cannot be consistent with any kind of deliberate creation. That notion has been painfully shed. At the moment I can't find any rational argument to knock down the view that argues for conversion to God . . . Now we realize the only logical answer to life is creation—and not accidental random shuffling" (in Seckbach and Gordon, 2009, pp. 343-344). The more scientists allow themselves to break free of their materialism to ponder deep metaphysical questions the more likely they are to come to the same conclusion.

The RNA World Hypothesis

We have noted that there are many theories of abiogenesis, but two have emerged as the major contenders: the RNA world and metabolism-first hypotheses. Various components of RNA have been synthesized in the lab, which requires intelligent chemists to control reaction rates and to know and select the right component in the right order for each reaction. Because they have done it, they assume that given enough time nature can do the same thing. Unlike chemists, however, nature is not intelligent and cannot think ahead, but it would nevertheless require the same control, order, selectivity, and knowing what the end product should be. Crucially, nature would have to keep the bits and pieces that are posited to end up as an RNA molecule away from water because water is "inherently toxic to polymers (e.g., RNA) necessary for life." Even the RNA monomers (the bases) have problems because water transforms them into different monomers, thus destroying the information carried by RNA (Neveu, Kim, and Benner, 2013, p. 394).

DNA, RNA, and proteins work as a unit, with DNA storing information, RNA reading and conveying it, and proteins doing the necessary enzymatic work. DNA requires enzymes (proteins) to replicate, but enzymes can only be synthesized by DNA; neither can exist without the other. To get this going, L-amino acids would have to have lined up just right to produce the protein at the same time and at the same place that nucleic acids arranged themselves to make RNA. Then these molecules must join hands to form a functioning, inseparable, irreducible whole. The probability of this occurring without intelligent guidance is almost beyond calculation. We know that the DNA/RNA/protein system is far too complex to have arrived spontaneously as a system, so which came first? Because RNA can store genetic information and self-replicate like DNA and perform the required enzymatic activity of proteins, many thought the RNA-first hypothesis would solve the chicken-or-egg problem. However, Harold Bernhardt points out that RNA is inherently unstable, and that the best ribozyme replicase (a molecule that catalyzes its own replication) created so far in the lab is about 190 nucleotides in length: "far too long a sequence to have arisen through any conceivable process of random assembly." He notes that it requires between 10^{14} and 10^{16} randomized RNA molecules "as a starting point for the isolation of ribozymic and/or binding activity in *in vitro* selection experiments, completely divorced from the probable prebiotic situation" (Bernhardt, 2012, p. 7).

The self-replication of RNA is the cornerstone of the RNA hypothesis, but no one has been able to achieve this in the lab. The enzyme/replicator double-duty presents us with a paradox because the two roles require contradictory properties. An enzyme, being a protein, must fold and be reactive or it is useless, while an RNA molecule carrying information must not do either or it would lose information. Moreover, the efficiency and fidelity of replication must be sufficient to produce viable copies at a rate exceeding the rate of decomposition of the parent molecule, and RNA is inherently unstable. Robertson and Joyce call it a myth that "a small RNA molecule that arises de novo and can replicate efficiently and with high fidelity under plausible prebiotic conditions. Not only is such a notion unrealistic in light of current understanding of prebiotic chemistry, but it should strain the credulity of even an optimist's view of RNA's catalytic potential" (Robertson and Joyce, 2012, p. 7).

The premise of OoL research is that life is just a matter of getting the physics and chemistry right, after which biology will take over. While life must be consistent with the laws of physics and chemistry it cannot be derived from them. The incredible complexity of life runs on the information content of DNA, but there is no information without an interpreter. Just as the information on a DVD disk needs a DVD player to convert the tracks into images and sounds, the information contained in the genes must have the cellular machinery to transcribe the message into a protein. One without the other is useless because they function as a unit. Andrew McIntosh writes of the irreducible complexity involved in creating all the connections: "All of these functioning parts are needed to make the basic forms of living cells to work...It is against the known principles of thermodynamics in physics and chemistry for this to happen spontaneously" (McIntosh, 2009, p. 370). Even granting the existence of self-replicating RNA surrounded by all the right L-amino acids, it cannot make a protein unaided by all the other necessary cellular components since RNA contains only raw information. A blueprint cannot make anything without the "workers" in the cell that understand it and can assemble what it codes for.

The Metabolism-First Hypothesis

Metabolism is the mechanism by which living things circumvent the second law by harvesting outside energy (food, water, sunlight) from the environment. Metabolism refers to all chemical processes (millions of chemical reactions) that occur in your cells that enable all living things to grow and thrive. It converts the food into energy to fuel cellular processes, such as building proteins and nucleic acids, and a method of eliminating cellular waste. David Abel notes: "Metabolism is the most highly integrated, holistic, conglomerate of organized formal functions known to science. How did life get so organized and goal-oriented out of an inanimate prebiotic environment that could care less about function or useful work? Chance and necessity cannot pursue function, let alone such an extraordinary degree of cooperative work" (Abel, 2011, p. 123).

To do such work there must be a lipid membrane boundary between the cell and the outside world. This membrane is far from a simple sac holding together the contents of the cell. It is a double-layered lipid/protein membrane of great complexity. The membrane acts as castle walls with multiple drawbridges that selectively allow the entry of vital resources required by the castle's residents and the exit of things needed outside the walls. This functional specialization is called compartmentalization. The importance of compartmentalization implies that the cell would have to have come before metabolism: After all, what is the point of metabolism unless you have a compartmentalized cell to sustain it?

The metabolism-first model is animated by the problems of the RNA world model and proposes the spontaneous formation of simple molecules, such as the compound formed from carbon dioxide and water, which triggered life. Such a primitive cell is assumed to have contained proteins (ignoring the difficulties involved in making them) possessing a crude non-genomic replication capacity (whatever that could be), and subsequent evolution processes somehow led to the accumulation of simple organic molecules that could serve as catalysts for more complex molecules containing RNA and DNA. Organic chemist Addy Prost says that the metabolism first model runs up against that pesky second law again: "How would metabolic cycles form spontaneously from simple molecular entities, and more importantly, how would they maintain themselves over time? We run yet again into that thermodynamic brick wall" (Pross, 2012, p. 107).

Research has shown that the metabolic systems proposed by metabolism-first proponents are unable to retain information about their composition to allow them to evolve toward a metabolic pathway. That is, they do not contain hereditary information by which they could pass on their composition to progeny. Commenting on both the RNA and metabolism first scenarios, Vasas, Szathmáry, and Santos maintain that: "Both schools acknowledge that a critical requirement for primitive evolvable systems (in the Darwinian sense) is to solve the problems of information storage and reliable information transmission" (Vasas, Szathmáry and Santos, 2010, p. 1470). These problems have led some to argue for top-down causation in the form of abstract information: "the key distinction between the origin of life and other 'emergent' transitions is the onset of distributed information control, enabling context-dependent causation, where an abstract and non-physical systemic entity (algorithmic information) effectively becomes a causal agent capable of manipulating its material substrate" (Walker and Davies, 2013, p. 7).

Information: The Recipe for Life

A third hypothesis for the OoL is the "information first." Arguing that information holds the key to the mystery of life's nature and origin, Sara Walker and Paul Davies state that "Although it is notoriously hard to identify precisely what makes life so distinctive and remarkable there is general agreement that its informational aspect is one key property, and perhaps the key property. The manner in which information flows through and between cells and sub-cellular structures is quite unlike anything else observed in nature" (2013, p. 1). Walker and Davies see life as emerging from a transition from bottom-up reductionist chemistry to top-down information flow and management in the causal structure: "The origin of life may thus be identified when information gains top-down causal efficacy over the matter that instantiates it" (Walker and Davies, 2016, p. 8).

But information is an abstraction, and abstractions are not usually seen as causal agents. Energy is also abstract, and we accept energy as a causal factor. Although energy is defined as the ability of a physical system to do work on another, we *infer* its existence by its effects, but we do not know what the essence of energy *is*. Information is like this in biology; it transfers knowledge of what to do from one living system to another without us being able to say specifically what it is. Information is indeed abstract and is always created by intelligence and thus by a mind. Only a mind of infinite wisdom could create the information necessary to make life. In a later paper, Walker and Davies appear to agree. They developed a model in which they talk of the fine-tuning of information and note that if the pathway from chemistry to life is the result of "fixed dynamical laws, then (our analysis suggests) those laws must be selected with extraordinary care and precision, which is tantamount to intelligent design: it states that 'life' is 'written into' the laws of physics *ab initio* ['from the beginning']. There is no evidence at all that the actual known laws of physics possess this almost miraculous property" (Walker and Davies, 2016, pp. 5-6).

Every living thing is a system composed of many thousands of separate parts that are functionally interdependent. All molecules and cells in an organism are in an information-rich cooperative relationship with all other molecules and cells by sending and receiving *information* on which they must act or else it all breaks down. Information transfer can only occur when both sender and receiver are "intelligent" enough to know what the information entails. Information may be described as symbolically encoded messages for carrying out a specific task or eliciting a specific response. The DNA/RNA protein system is a set of recognizable grammatical symbols that have been taken and put together as words to construct a syntactically correct sentence. DNA is a natural code that possesses a set of abstract symbols (the base letters: AGCT) with syntactic rules. Sets of letters do not by themselves necessarily convey meaning; meaning is determined by the natural language of a system. Physicist Hubert Yockey points out that the phrase: "'O singe fort' has no meaning in English, although each is an English word, yet in German, it means 'O sing on,' and in French, it means 'O strong monkey'" (Yockey, 2005, p. 6). DNA is a specific language with its own meanings, just as English, French, and German are.

To mean something in the language of DNA the sequence of nucleotides must have explicit specificity; the sequence of letters must form a syntactically correct sentence. RNA contains instructions for specific amino acids and their sequence, and those instructions require a reply on the part of the receiver. When ribosomes read the instructions from the messenger RNA, they respond by forming bonds between specific amino acids to make the specified protein. The totality of these processes results in a living, functioning organism like you and me. Werner Gitt observes that the question of "How did life originate?" is inextricably linked to the question "Where did the information contained in all those base sequences in the genetic code come from?" (Gitt, Compton, and Fernandez, 2011, p. 169). John Lennox is not surprised that abstract information is fundamental to life, and states: "This proposal, that information be regarded as a fundamental quantity, has profound implications for our understanding of the universe. But it is not new, it has been around for centuries. 'In the beginning was the Word...all were made by Him'" (2009, p. 177).

The optimism after the 1953 Miller-Urey experiment that it would be relatively easy to kick-start life in the lab has slowly faded to pessimism. Even Urey admitted that while he believes in abiogenesis, he does not do so by dint of evidence, but by *faith*: "All of us who study the origin of life find that the more we look into it, the more we feel that it is too complex to have evolved anywhere. But we believe as an article of faith that life evolved from dead matter on this planet. It is just that its complexity is so great, that it is hard for us to imagine that it did" (in Persaud, 2007, p. 84). Science has gained an immense amount of chemical and biological knowledge in its search for the OoL, and that is a big plus. Many thousands of biologists and chemists have spent millions of hours experimenting and calculating since Miller-Urey, and this has resulted in a clearer understanding of the immensity of the problem rather than its solution. This does not mean that scientists must stop trying to discover natural explanations for the OoL. Regardless of an individual scientist's religious convictions, he or she cannot stop and conclude that God did it. We know that He did by "devising the rules of the game," as Nobel laureate physicist Erwin Schrodinger said. But Schrodinger also said that God left it up to science "to discover or to deduce" (in Moore, 2015, p. 348).

Chapter Eleven
DNA: God's Book of Life

Decoding the Book of Life

The genome is God's construction manual that provides the information needed to build proteins, the complicated structures that build us. Scientists have struggled to unlock the secrets of the genome ever since the Augustinian abbot Gregor Johann Mendel established the early rules of heredity in the mid-eighteenth century. Progress was slow but steady, punctuated by some major advances such as the discovery of DNA in the 1870s, the discovery of the architecture of DNA in the 1950s, and DNA "fingerprinting" in the 1980s. This was capped in 2000 with the completion of the $2.7 billion Human Genome Project (HGP) that succeeded in sequencing the entire human genome. At the ceremony honoring this scientific feat, President Bill Clinton remarked:

> Today's announcement represents more than just an epoch-making triumph of science and reason. After all, when Galileo discovered he could use the tools of mathematics and mechanics to understand the motion of celestial bodies, he felt, in the words of one eminent researcher, that he had learned the language in which God created the universe. Today we are learning the language in which God created life. We are gaining ever more awe for the complexity, the beauty, the wonder of God's most divine and sacred gift. (Clinton, 2000)

The HGP's task was to identify and map all the genes of the human genome. The pooled wisdom of more than a thousand scientists from diverse disciplines from six nations was required to do this, and the task of plumbing its mysteries is still ongoing. Secular scientists believe that this exquisitely designed genome that required such brainpower to investigate arose fortuitously from ancient sludge by atoms just bumping into each other in the night. We wouldn't expect such a miracle of a simple instruction manual on how to assemble a child's bicycle, never mind one on how to assemble a human being. The genetic code has immense information content that reads, interprets, and edits itself; no other code comes remotely close to being able to do this. Scientists Gitt, Compton, and Fernandez tell us that: "From an engineering point of view, and under the criteria that were considered here, the code system used in living organisms for protein synthesis—the *Quaternary Triplet Code*—is the best of all possible codes considering the four requirements that must be met. This testifies to purposeful design" (Gitt, Compton, and Fernandez, 2011, p. 166).

The human genome is remarkably compelling biological evidence, second only to the intricacies of the human brain, of the existence of a Dive Designer that we have. Francis Collins, former head of the Human Genome Project and a former atheist, calls DNA the "language of God," and offers its mind-boggling complexity as a compelling argument for God. Gitt and his colleagues quote Romans 1:20 ("For since the creation of the world God's invisible qualities—his eternal power and divine nature—have been clearly seen, being understood from what has been made, so that people are without excuse") and argue that there is no excuse for denying God and his power because the evidence is just too compelling. Some will not think to contemplate the evidence, and some may not be able to, so they may be forgiven. But those who contemplate and still reject God will do so willfully.

Genes and Protein Making

We have trillions of cells in our bodies that, with some exceptions, are factories for making proteins. Thousands of proteins are constantly being made, and the information needed to make them is carried on specific segments of DNA called genes. DNA consists of two strings of nucleotides tightly wrapped around a protein core called a histone and twisted around each other to form a double helix ladder. Each nucleotide is made from a sugar and phosphate backbone, and a base (the ladder rungs). There are four different bases: adenine (A), thymine (T), cytosine (C), and guanine (G), that bond in specific ways: C can only pair with G, and A can only pair with T. Genes are the blueprints or recipes for life that contain all the information for the extraordinarily complex process of instructing cells what proteins to make and when to do it. A gene is a segment of DNA that codes for the manufacture of a specific protein. We inherit two forms of a gene—one from each parent—called *alleles* located at the same place on a chromosome. Alleles help to facilitate our different traits and behaviors by coding for different levels of a protein product. If more than one allele occupies a gene's locus in a population, the gene is said to be polymorphic. A polymorphic gene may have only one nucleotide substitution (underlined) in thousands, as in the example below, known as a single-nucleotide polymorphism (SNP). The bolded bases are known as a codon.

        Allele 1.   TCACCTTGGA**A**TGGGCTA

        Allele 2.   TCACCTTGGA**GT**GGGCTA

There are about 3 billion base pairs in the human genome, and a gene is a group of adjacent base pairs that code for a protein. While there are approximately 21,000 genes in the human genome, only about two percent of our DNA codes for proteins; many other segments regulate the behavior of the coding DNA. DNA's information content is enormous. The typical cell nucleus measures

11,811/50,000,000 of an inch (that's the simplest fraction!) and there are 6 feet of DNA packed into it. If the human body's entire DNA were to be unwound and placed end-to-end it would reach the sun and back (Clark and Pazdernik, 2009, p. 239).

A gene product may be a neurotransmitter such as serotonin that influences how we behave or feel, but they do not *cause* us to behave or feel one way or another, they *facilitate* our behavior and feelings. The relevant protein products produce *tendencies* or dispositions to respond to the environment in one way rather than in another; they do not determine those responses. God would not design a life-giving system that determined human behavior because He endowed us free will. As is the case with cosmology, the more scientists discover how exquisitely fine-tuned the amazingly complex human genome is, the more they stand in awe.

Our genomes are tuned to respond to our needs. Afferent nerves are a set of physiological thermostats that sense and transmit information by carrying nerve impulses from sensory organs to the brain about the state of our internal or external environments. When we require a specific protein, the nucleus kicks on and an enzyme called DNA helicase unzips the double-stranded DNA into two single strands. An enzyme called RNA polymerase (RNAP) then binds to the promoter region of a gene to signal the DNA to unwind so the bases on the DNA strand can be read to make a strand of messenger RNA (mRNA). This process of copying the instructions on the DNA to make a protein is called transcription. Uracil is substituted for thymine as the base that complements adenine at this time. When RNAP reaches a gene's stop sequence, the mRNA strand is complete, and it detaches from the DNA strand. The DNA double helix is then reconstituted by the billions of free-floating nucleotides in the nucleus and the mRNA begins its journey to the protein factory where the message is translated, and the specified protein made. DNA polymerases add nucleotides to a growing DNA strand at an astounding 50 nucleotides per second. Because the environment in the nucleus differs from that of the cell's cytoplasm, they are walled off by a double membrane with channels allowing for the passage of mRNA called the nuclear pore complex that recognizes and controls information flow. The top of figure 11.1 illustrates transcription and the bottom half illustrates translation.

Protein-making instructions are transmitted by mRNA in the form of a triplet of bases (e.g., CAA, AGC, CCU, etc.) called codons. Codons are four bases conveyed in units of three, so there are $4 \times 4 \times 4 = 64$ possible arrangements of them; more than enough for the coding of the 20 standard amino acids found in the body. Codons are three-letter words that correspond to the word for a particular amino acid, and a sequence of codons may be thought of as a legible sentence that is to be read by transfer RNA (tRNA). Transfer RNA picks up and transports the appropriate set of amino acids (anticodons) that complement the codons on the mRNA strand. Codon and anticodon are then slotted into place by yet another form of RNA called ribosomal RNA (rRNA), thus completing the protein chain. The process of changing information from the language of RNA into the language of amino acids is called translation. The completed protein is then sent from the cell to do its work by binding to a receptor protein in a body cell.

Translation completes the flow of genetic information; the next step is to get the amino acid chain to fold into the correct 3-dimensional shape it needs to fit into a receptor protein. biochemists Denton, Marshall, and Legge take us on a fascinating journey through the intricate process of protein folding, and inform us that: "It is more than anything else the complex hierarchic structure of the folds—their being composed of clearly defined substructures and submotifs combined together into what appear seemingly to be irregular complex hierarchic wholes, the sort of order which is so characteristic of that of a machine or artifact—which conveys the irresistible feeling that such forms *could not possibly be natural or lawful*" (Denton, Marshall, and Legge, 2002, p. 339). They note that the self-organizing structures folding within the cell are governed by a rich "vocabulary" of "words" (information). They do not tell us what is responsible for imparting that information in the cell, but we know that lifeless atoms cannot write their own software.

Every protein requires the right amino acids to function as intended, which requires genes to specify the correct sequence of amino acids before the folding can take place. There are four stages by which a chain of amino acids become are folded to become a functioning protein primary, secondary, tertiary, and quarternary; geek-speak for first, second, third, and fourth. The primary structure is the sequence of amino acids held together by their peptide bonds. The secondary structure is the protein beginning to fold up via various types of hydrogen bonds. Each amino acid interacts with the others and they either twist into a corkscrew-like helix or take the shape of a folded sheet. Proteins called chaperones act as catalysts facilitating the correct assembly but do not constitute a part of the assembled product. No cell can survive without chaperones, but since chaperones are proteins, what facilitated their correct folding before chaperones existed? During the tertiary stage, the protein is folded into its precise 3D structure specific to its function. This requires a number of forces working to produce interactions of groups of amino acids, with thermodynamics providing the stabilizing force on the protein by adjusting it to its stable state because the protein is fighting nature's tendency toward disorder. In the quaternary stage, a number of amino acid chains from the tertiary structures fold together into a global structure. Proteins must

DNA: God's Book of Life 107

contain the information to proceed to the only place where they can connect; that is, a receptor whose shape complements theirs like a lock and key.

Figure 11.1. The Making of a Protein

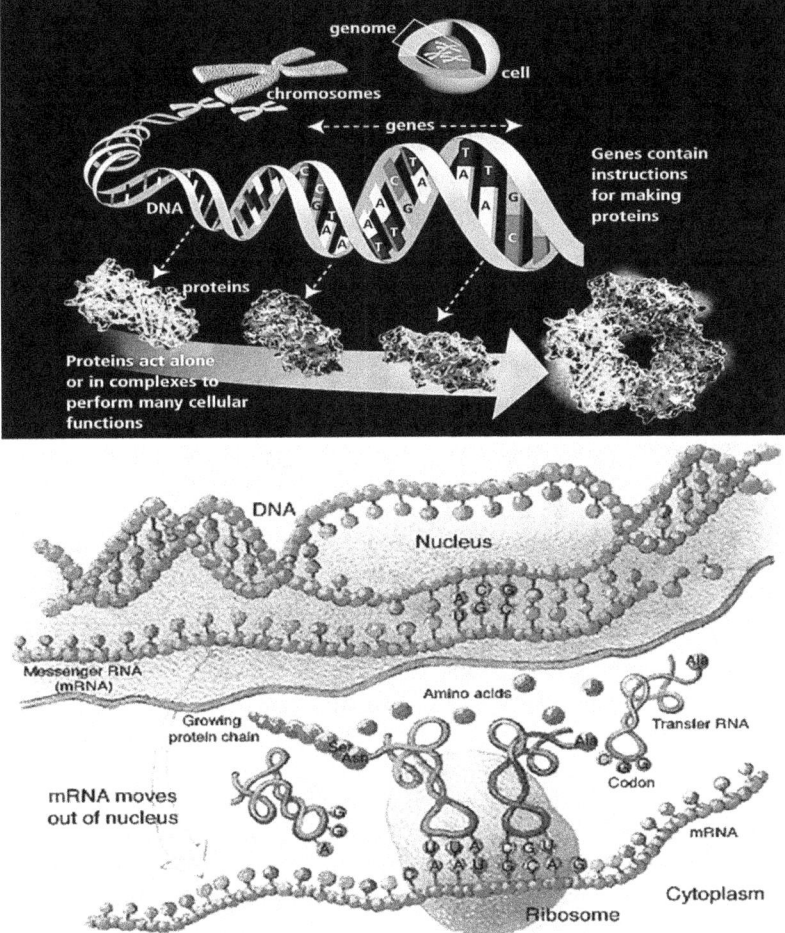

Allen and Lidström call the protein folding process a "profound mystery of seemingly impossible complexity," and say that it is "far beyond the ability of current computer simulations to replicate" (Allen and Lidström, 2016, p. 36). However, efforts have been made to do so, including one using an IBM computer with tremendous computing power called Blue Gene. IBM scientists noted that protein folding holds the key to understanding the basics of life: "The scientific community considers protein folding one of the most significant 'grand challenges'—a fundamental problem in science or engineering." They also stated that: "Blue Gene's massive computing power will initially be used to model the folding of human proteins, making this fundamental study of biology the company's first computing 'grand challenge'" (IBM, 1999). A senior IBM computational biologist stated that if everything goes as planned it will take "Blue Gene about a year to simulate on the computer the folding of a single protein. How long does it take the body to fold one? Less than a second. It is absolutely amazing the complexity of the problem and the simplicity with which the body does it every day" (Lohr, 1999). If great scientific minds built this remarkable computer capable of more than 1,000 trillion operations per second and takes a year to merely simulate what our genomes do thousands of times each day in less than a second, how great is the Mind that built us?

A major argument against a Designer of life used by secular biologists used to be what they considered to be "junk" DNA in the Genome. It was so-called because 98% of our DNA does not code for proteins, and these non-coding regions were thought of as DNA "fossils" that were once functional but now are not. It was argued that no designer worth his salt would leave junk floating around doing nothing in his creation. A decade-long project called Encyclopedia of DNA Elements (ENCODE) ended such thinking. ENCODE is a consortium of 442 researchers from 32 different institutions around the world and is a follow-up to the Human Genome Project. ENCODE looks at the non-coding parts of DNA, which is a more daunting study than the coding parts. After five years of lab work and the equivalent of 300 years of computer time, ENCODE scientists found that 80% of the human genome serves some biochemical function in regulating when, how, and where a gene is activated, with the promise of new technology finding function for the remainder. The project has identified about 10,000 stretches of DNA for making the RNA molecules required to regulate the actions of the protein-coding genes. As Stephen Hall remarked: "The ENCODE project has revealed a landscape that is absolutely teeming with important genetic elements—a landscape that used to be dismissed as 'junk DNA'" (Hall, 2012, np). ENCODE's senior researcher Ewan Birney said: "By carefully piecing together a simply staggering variety of data, we've shown that the human genome is simply alive with switches, turning our genes on and off and controlling when and where proteins are produced" (National Institute of Heath, 2012, np).

Geneticist Nessa Carey scorns the notion of calling non-coding DNA "junk" because it is not responsible for coding. She imagines a car factory in which only two people (the 2% of protein-coding genes) were involved in building a car and 98% percent (the percentage of non-coding DNA) sitting around idly doing nothing. It is ridiculous to think that only two people are needed to run the factory, and it is ridiculous to think this way about the genome. Carey says that cars are the endpoint of an automobile factory in the same way that proteins are the endpoint of DNA, but neither could be produced without the "junk." Just as two people cannot sustain a successful car brand, neither can the other 98 if no cars are made to sell: "The whole organization only works when all the components are in place. And so it is with our genomes" (Carey, 2015, p. 3). The genome is an irreducibly complex molecular system of multiple interdependent parts that requires all components to be in place for the system to function.

As surprised as geneticists were with ENCODE's findings, from an anthropic point of view their results were to be expected. In 1998—before the Human Genome Project was completed and before ENCODE—intelligent design proponent William Dembski predicted that "junk" genes had a function: "On an evolutionary view we expect a lot of useless DNA. If, on the other hand, organisms are designed, we expect DNA, as much as possible, to exhibit function" (in Meyer, 2009, p. 407). Dembski made this prediction on the basis of simple anthropic design logic; that is, an information processing system such as the genome cannot function if it contains mostly useless parts.

The Wonders of the Human Cell

Before the invention of the electron microscope, living cells were thought of as simple blobs of protoplasm gluing us together. We now view them as a collection of super-efficient factories containing thousands of living entities all working in tandem to provide for our needs. Moreover, these living power plants constantly divide without losing cohesiveness. Cells are three-dimensional living structures formed from a one-dimensional string of instructions contained in the genome. There are over 200 different kinds of cells in our body that perform different tasks; some are brain cells, others make bone, muscle, and hair, and others make red or white blood cells. They are short-lived creatures, though, with about 300 million dying every minute in a form of cellular suicide known as apoptosis, or programmed cell death. This is a normal process in which enzymes destroy the DNA in the cell's nucleus to make room for fresh ones.

Everything discussed in this chapter so far takes place in the cell. The information in DNA must be stored, transcribed, and translated, and then the output of all this activity must be inspected, packaged, and sent to its proper destination. This activity requires an awful lot of hardware and software packaged in this marvel of precision atomic engineering that could be fitted into a space smaller than the period at the end of this sentence. It is impossible to do justice to all the work the cell does here, but too much detail takes the mind off the "big picture" anyway. It is the big take-away picture I want to paint, and that is a sense of wonder at the marvelous design and stunning complexity of the cell. Bill Bryson captures it well in four sentences when he writes:

> Every cell in nature is a thing of wonder. Even the simplest are far beyond the limits of human ingenuity. To build the most basic yeast cell, for example, you would have to miniaturize about the same number of

components as are found in a Boeing 777 jetliner and fit them into a sphere just 5 microns [0.00019685 inches] across; then somehow you would have to persuade that sphere to reproduce. ... But yeast cells are nothing compared with human cells, which are not just more varied and complicated, but vastly more fascinating because of their complex interactions. (2003, p. 372)

Each one of our multiple trillions of cells is an incredibly complex and organized marvel of biological nanotechnology designed to work towards the whole organism alive and kicking for three-score years and ten. To accomplish this marvelous feat requires that the cell work on a myriad of subordinate molecular goals that somehow collectively "know" what those ultimate goals are. The cell is a hive of non-stop chemical activity of mind-boggling complexity. It has been likened to a super-factory with everything needed to supervise, plan, construct, package, and transport the protein products we need to live contained within. Unlike brick-and-mortar factories, cells continually make copies of themselves over their lifespan and are unimaginably more complex.

As is the case with any factory, the cell has many structures required to fulfill its purpose. We start with the cytoskeleton, which is the structural foundation of the cell that determines its shape, just as the skeletal bones give shape to the body. It also constitutes the assembly conduit directing the organelles and other substances around the cell. The nucleus is the control center from which the DNA boss sends out instructions about what proteins are needed at the time, and ribosomes are the workers that build them on the assembly line. These workers, working at warp speed, have to arrange the right amino acids in just the right sequence every time out of the millions of possible ways they can be arranged. The Golgi apparatus bundles proteins as they are synthesized, and the molecules that form the cell's membrane serve as security guards that monitor which substances are allowed in and out of the cell through which entrance or exit. The cell membrane is a layered marvel of design that carries out other functions as well. The factory floor is the cytoplasm, and the vacuoles are a kind of warehouse that stores the nutrients a cell needs as well as storing waste products awaiting disposal, thus protecting the cell from contamination. The endoplasmic reticulum is part of the quality control mechanism that inspects the finished product to ensure that only correctly folded proteins are sent to their destinations, and the lysosomes that break down waste in the cell and discard it are the janitorial staff.

On rare occasions, proteins are folded the wrong way. To counter this, the cell has a quality-control system for proofreading newly synthesized proteins. If they are defective, they are either repaired or stripped down to their component parts. DNA thus contains specific information that results in the building of all kinds of living things with astounding precision. Errors in the protein folding process not caught by the cell's elaborate surveillance system occur at about one error per 10 billion nucleotides and are the mutations required for natural selection.

All systems require energy to function, and cells get theirs from over 1,000 organelles per cell called mitochondria. Mitochondria use cellular oxygen to convert chemical energy from food in the cell into adenosine triphosphate (ATP). Mitochondria have their own separate genomes and multiply automatically when their cells need more energy. ATP's energy is stored in chemical bonds which can be opened and the energy redeemed. It has been noted that: "The efficiency of the transport of electrons in the last stage of ATP synthesis is 91%, an efficiency of which engineers can only dream" (Gitt, Compton, and Fernandez, 2011, p. 297). The cell is a perfect example of what scientists call irreducible complexity; that is, a complex system that requires all its parts to be in place for it to function, and that if any of the interacting parts were removed the entire system becomes non-functional.

Cells are thus systems within systems within systems, and it seems logical that would have to have arrived on the scene as a whole unit to be functional. Cars can't assemble themselves without workers, or proteins without ribosomes; and where was the ATP to energize the process? What if it needed repair and there was no repair mechanism? If the repair mechanism came first, what would be the point if there was nothing to repair? How likely is it that the human genome's 3 billion-plus letters arranged in an orderly fashion, and conveying more information than a huge library of books "just happened" rather than the product of a mind with infinite intelligence?

It was this complexity that led the most notorious atheist of the twentieth century, British philosopher Anthony Flew, to come to believe in God. Flew noted that his conversion resulted from the increasing number of physicists who, with Fred Hoyle, see a super-intelligence behind the anthropic universe, and the amazing complexity of DNA: "I believe that the origin of life and reproduction simply cannot be explained from a biological standpoint despite numerous efforts to do so. With every passing year, the more that was discovered about the richness and inherent intelligence of life, the less it seemed likely that a chemical soup could magically generate the genetic code" (in Walsh, 2020, p. 127). Another famous British former atheist is biophysicist Allister McGrath. While studying for his Ph.D. at Oxford University, he wrote: "I was discovering that Christianity was far more intellectually robust than I had ever imagined. I had some major rethinking to do, and by the end of November [1971], my decision was made: I turned my back on one faith and embraced another" (2010, p. 81). McGrath's former faith was Marxism; Flew's was socialism, and both of these pernicious faiths are Siamese-twinned to atheism and tyranny.

CHAPTER TWELVE

Evolution by Natural Selection: Micro and Macro

Charles Darwin and his Dissenters

Biologist Bernard Wood notes there are images everywhere drawing a straight line from fish, to retile, to ape, to a graceful human striding toward the future, and writes: "Our progress from ape to human looks so smooth, so tidy. It's such a beguiling image that even the experts are loath to let it go. But it is an illusion" (Wood, 2002, p. 44). I am not pushing an anti-evolution agenda, but many scientists have some serious problems with it. Thousands of them have signed the *Scientific Dissent from Darwinism*, which reads: "We are skeptical of claims for the ability of random mutation and natural selection to account for the complexity of life. Careful examination of the evidence for Darwinian theory should be encouraged." A lot more would sign a statement affirming their belief in Darwinism, but headcounts do not settle scientific issues. I mention the *Dissent* only to show that Darwinism is not in the same scientific league as theories in physics and chemistry.

In his *Mathematics of Evolution*, Fred Hoyle explains mathematically why many Darwinian claims are outside the realm of possibility. He did not dispute microevolution (small changes within a kind); his argument was with macroevolution (speciation). He argued that species can only adapt within the narrow limits of their kind and that the extrapolation from micro to macroevolution has led us into a deep scientific bog. Likewise, Nobel laureate physicist Robert Laughlin calls macroevolution theory an "anti-theory" that prevents rather than stimulates thinking. He writes: "Your protein defies the laws of mass action? Evolution did it! Your complicated mess of chemical reactions turns into a chicken? Evolution! The human brain works on logical principles no computer can emulate? Evolution is the cause!" (Laughlin, 2005, pp. 168-169). What is this theory that causes such disturbance, and what of its author, Charles Darwin, whom Richard Dawkins credits with making it intellectually respectable to be an atheist?

Darwin never called himself an atheist and was a firm believer in God, as the following passage, written 33 years after the publication of his *The Origin of Species* shows:

Another source of conviction in the existence of God, connected with the reason and not with the feelings, impresses me as having much more weight. This follows from the extreme difficulty or rather impossibility of conceiving this immense and wonderful universe, including man with his capacity of looking far backwards and far into futurity, as the result of blind chance or necessity. When thus reflecting I feel compelled to look to a First Cause having an intelligent mind in some degree analogous to that of man; and I deserve to be called a Theist. (Darwin, 1892, p. 61)

Darwin denied that his theory leads to atheism "It seems to me absurd to doubt that a man may be an ardent Theist and an evolutionist." Darwin said that his level of belief fluctuates, but: "In my most extreme fluctuations I have never been an atheist in the sense of denying the existence of a God" (Darwin, 1879). Darwin had great confidence in his theory of natural selection, but he also believed that the laws of nature that allowed for evolution stemmed from God. He wrote in *The Origin:* "To my mind, it accords better with what we know of the laws impressed on matter by the Creator, that the production and extinction of the past and present inhabitants of the world should have been due to secondary causes, like those determining the birth and death of the individual" (Darwin, 1982, p. 458). Secondary causes are the natural properties of matter granted power by the Creator to produce changes themselves without requiring His micromanagement, but subordinate to the primary causes immanent in the laws of nature from the beginning of the universe.

The atheist view of natural selection is that it is devoid of purpose: "Natural selection, the blind, unconscious, automatic process... which we now know is the explanation for the existence and apparently purposeful form of life, has no purpose in mind. It has no mind and no mind's eye. It does not plan for the future. It has no vision, no foresight, no sight at all" (Dawkins, 2006, p. 9). Darwin, however, saw purpose in evolution. As he put it in *The Origin:* "Hence we may look with some confidence to a secure future of equal inappreciable length. And as natural selection solely by and for the good of each being, all corporeal and mental endowments will tend to progress towards perfection" (Darwin, 1982, p. 459). In an 1881 letter to a friend, Darwin wrote: "If we consider the whole universe, the mind refuses to look at it as the outcome of chance—that is, without design or purpose" (Darwin, 1982, p. 459).

Evolution occurs via the selection of favorable genetic variants that increase the "fitness" (a quantitative evaluation of a phenotype gauged by the number of its progeny relative to other phenotypes in a breeding population). The basic idea is that populations of living things grow until they strain the ability of the environment to support all members, resulting in a struggle for existence in which only the fittest survive. Individuals within populations exhibit variation with respect to phenotype (disease resistance, size, speed, cunning, etc.) that aid in the struggle. Thus, there must be heritable trait variation in a breeding population that confers an edge on some in the struggle for survival in a prevailing environment. Organisms with traits better suited to their environment will reproduce more and thus increase the proportion of the population with those traits. The arrival of an advantageous trait is the result of a genetic mutation the cell's repair mechanism failed to catch. Most mutations are neutral, but many others are harmful in that they reduce the fitness of an organism. However, if a beneficial mutation arises that increases its carriers' reproductive success, and if it proves sufficiently advantageous, the allele underlying the trait will arrive at "fixation." Fixation occurs when an advantageous mutant allele arises in a population and completely replaces all other alleles after a certain number of generations.

Necessity and Information

Darwinists aver that the accumulation of small quantitative genetic changes *within a species* eventually results in large qualitative changes and a totally new species (macroevolution) and look to chance and necessity to explain it. The chance half is the random occurrence of a mutant allele in a mating population that just happens to be advantageous in a particular environment at a particular time. The necessity half is the process of natural selection that generates order in the genome by preserving the useful and eliminating the harmful. Without this winnowing process, mutations would yield only disorganization and extinction because of the excess of harmful mutations. Necessity implies that something is predestined and could not be otherwise, such as the fact that apples always fall downwards from the tree.

Necessity is based on the belief that life issues from the laws of chemistry and physics work in nature to produce new species. However, George Williams notes that evolutionary biologists fail to realize that they work with two incommensurable domains: information and matter. Genes are packages of information, and although DNA is the material medium that specifies a gene, it is not the message (Williams, 1992, p. 11). The three-letter code for the amino acid arginine is AGC, but there are no laws of physics or chemistry that say those three letters must code for that particular amino acid. The laws that govern the physical and chemical properties of DNA do not determine the sequence on a string of DNA: "If it was so, the string could not contain any information. For DNA to work as a carrier of genetic information, it was necessary that this molecule acquire the capability to change its sequence arbitrarily...there is nothing from chemistry or physics that can be used to derive the function of DNA. This function is irreducible" (Bauchau, 2006, p. 36).

Hubert Yockey likewise informs us that the principles of biology are not reducible to the laws of physics and chemistry because even to construct the simplest organism requires more genetic information "than the information content of these laws. The existence of a genome and the genetic code divides living organisms from nonliving matter. There is nothing in the physicochemical world that remotely resembles reactions being determined by a sequence and codes between sequences" (Yockey, 2005, p. 2). Thus, we see two remarkable things about the DNA code: (1) It has the ability to change itself in response to environmental conditions—this is a necessary requirement for natural selection, and (2) the information content of the genome is far greater than the information content of physicochemical laws. The DNA code is *information,* and information is non-material. It depends on matter and energy for storage and movement, but it is not defined by the biochemistry of the molecules used for these purposes. Rather, it is the information that defines the operations of the matter. DNA is a molecule beyond what we normally think of as a chemical model: "Was it perhaps the power, thinking, and will of a supreme being that created this self-replicating basis for life... How do we define something that can stretch, split along its middle, and clone itself—as a chemical molecule or as a living thing?" (Aczel, 1998, p. 88).

DNA is unfathomably more complex than a single-direction code like a computer's binary code. It is bidirectional in that it relays different messages when read in opposite directions so that multiple sets of instructions can be embedded in DNA overlap. The nucleotide sequence on the end of one gene overlaps a second sequence, and the letters of the first sequence have a different meaning than they have on the second. A sequence of nucleotides can thus code for more than one gene product by using different reading frames. Stephen Meyer makes an analogy with spy codes. A spy message read one way by an enemy might be about mundane goings-on down on the farm, but a friendly reader in possession of the key would read it as intended. Within the cell, the RNA, proteins, and enzymes work together to access, identify, and transcribe these messages within messages. Thus: "The presence of these genes imbedded within genes (messages within messages) further enhances the information-storage density of the genome and underscores how the genome is organized to enhance its capacity to store information" (Meyer, 2009, p. 463).

Geneticists have noted that the: "Maintenance of dual-coding regions is evolutionarily costly and their occurrence by chance is statistically improbable," and then say that dual overlapping protein-coding is "virtually impossible by chance" (Chung et al., 2007). Placing the adverb "virtually" before the adjective "impossible" allows them to avoid charges of advocating intelligent design, and to think that the codes somehow wrote themselves. Think of the brainpower needed to write a spy's encrypted message. The spy has to use standard language so that both friend and foe can understand it, and both the manifest and hidden messages must convey real meaning. If the manifest message did not it would arouse the enemy's suspicion, but if the encrypted message did not convey real meaning, it would be useless. Of course, there is no intention to deceive in the genome; Meyer's analogy is just his way of revealing the ingenuity of the code that allows its "readers" (the ribosomal molecules) to understand multiple messages backward and forward.

The ability of DNA to convey information relies on its freedom from chemical determinism. If it were not, its information conveying capacity would be destroyed. DNA's information content is no more reducible to the laws of physics or chemistry than the information of a newspaper originates from the chemistry of paper and ink: "Instead, the genetic code functions as a higher-level constraint distinct from the laws of physics and chemistry, much like a grammatical convention in a human language" (Meyer, 2009, p. 240). Meyer shows that no chemical bonds exist between the bases along the longitudinal axis of the helix where the genetic information is and that identical bonds link all the nucleotide bases to the sugar-phosphate backbone of the double-helix.

In a further analogy, Meyer points out that magnetic letters on a metal surface can be combined and recombined to form any sequence. The laws of physics determine that magnetic letters placed on a metal surface determine that they will stick, but there is no law of attraction that determines their arrangement into a meaningful sequence, just as there are no laws of chemistry that dictate the arrangement of DNA bases to form a meaningful biological sentence. The four bases attach to any site on the DNA backbone with equal facility because the same chemical bond occurs between any of the bases and the backbone; there is no chemical affinity for one rather than another that accounts for sequence variations. If chemical determinism governed the DNA molecule there would be only one kind. Deterministic natural laws—by definition—will always produce the same result, so the arrangement of DNA bases would be the same each time. That they are not points to the specified complexity of DNA.

Macroevolution and the Problem of Time

Biologist Gerd Muller informs us that microevolution has been confirmed and if our theorizing was confined to this level there would be no controversy. He chides evolutionists for habitually taking the success of small-scale evolution as the "explanation of *all* evolutionary phenomena," and points out that "a wealth of evolutionary phenomena remains excluded. For instance, the theory largely avoids the question of how the complex organizations of organismal structure, physiology, development or behavior—whose variation it describes—actually arise in evolution" (Müller, 2017, p. 3). The problem is that time is a huge impediment to macroevolution. Take enzymes for example. Experiments with directed mutations in enzymes with just seven nucleotide substitutions make a conversion estimate that it would take 10^{30} generations to get an enzyme with a new functional fold. This is a timescale way beyond life on Earth (Gauger and Axe, 2011, p. 13). In other words, new enzymes cannot be reconfigured through a gradual process of mutation and selection because 10^{30} generations, even if a generation lasted one second, would take far more time than all the seconds that have ticked by since the Big Bang. This is seconded by three microbiologists who note that if more than the tiniest fraction of amino acids is important for making the enzymes different, "then it may be effectively impossible for undirected mutations to stumble upon the right combinations for functional conversions… The problem for evolutionary explanations is that the very special circumstances needed to achieve even weak conversions in the lab translate into highly unrealistic evolutionary scenarios" (Reeves, Gauger, and Axe, 2014, pp. 10-11).

Mathematical models have shown that the "waiting time" to form a specified string of nucleotides in a hominid population of 10,000 individuals by mutation/selection under ideal conditions would require millions of positive mutations: "In small populations, the waiting time problem appears to be profound, and deserves very careful examination" (Sanford et al., 2015, p. 27). One OoL researcher calculated how long it would take for life to kick-start itself from amino acids randomly banging against one another to make a protein: "Imagine that every cubic quarter-inch of ocean in the world contains ten billion precellular ribosomes. Imagine that each ribosome produces proteins at ten trials per minute (about the speed that a working ribosome in a bacterial cell manufactures proteins). Even then, it would take about 10^{450} years to probably make one useful protein" (Klyce, nd). This is just the difficulty of the mutation and natural selection of one protein. How about the millions of mutations, with intermediary mutations more likely to be maladaptive than adaptive, required to go from sludge to Shakespeare? It becomes exponentially unlikely when we realize that to produce a new phenotypic trait such as an arm or an eye in by natural selection requires genetic innovation to control metabolic pathways, and such innovation requires countless *coordinated* sequences of enzymatic steps, not innovation in one isolated enzymatic function.

Genetic mutations are the holy grail of natural selection but experiments of selective breeding with thousands of generations of fruit flies produce nothing but weird fruit flies with features that are maladaptive rather than adaptive. Moreover, they quickly achieve genetic homeostasis; that is, their gene pool runs out of variation capacity. Studies of 30,000 generations of E. coli (the equivalent of about a million human years) produce only harmful *de*volutionary results, losing many of the building blocks of RNA, rather than beneficial evolutionary results. Bacteria have an easier shot at gaining favorable mutations than humans because they have exponentially larger populations and reproduce far more rapidly. Bacteria evolve to be resistant to antibiotics, but these are adaptations "within kind," and no new body parts have evolved to make them other than what they were when they first arrived on Earth about 3.5 billion years ago. Evolution requires adaptive mutations, not maladaptive ones, and the latter are many times more common. Elevated mutation rates help advantageous traits to spread through a population faster, but it also hurts by increasing mutation load and thus decreasing overall fitness.

Macroevolution, the Cambrian Explosion, and the Tree of Life

At the heart of macroevolution is *speciation*, the formation of new species. Species are genetically distinct groups of interbreeding animals that cannot reproduce with animals not of their kind. It is all but impossible to demonstrate speciation, although the National Academy of Sciences (NAS) says otherwise: "A particularly compelling example of speciation involves the 13 species of finches studied by Darwin on the Galápagos Islands, now known as Darwin's finches" (Ayala et al., 1999, p. 10). The fact that finches remained finches and can interbreed, notwithstanding, none of these so-called "species" are distinct; they simply vary in small morphological differences within their kind called ectomorphs. Thus, the NAS was perpetuating a fiction they knew to be false, which led Phillip Johnson to write, "When our leading scientists have to resort to the sort of distortion that would land a stock promoter in jail, you know they are in trouble" (Johnson, 1999).

If speciation is true, we should find many transitional fossils between species, but we do not. Nils Nilsson, an evolutionist at Lund University in Sweden, expresses his disappointment that his life's work for the search of transitional forms: "My attempts to demonstrate evolution by an experiment carried on for more than 40 years have completely failed... The fossil material is now so complete that it has been possible to construct new classes, and the lack of transitional series cannot be explained as being due to scarcity of material. The deficiencies are real, they will never be filled" (in Nitardy, 2012, p. 60). Stephen J. Gould admitted that it is a "trade secret" of paleontology that fossils that could plausibly be considered transitional forms are exceedingly rare, and adds: "The evolutionary trees that adorn our textbooks have data only at the tips and nodes of their branches ... in any local area, a species does not arise gradually by the gradual transformation of its ancestors; it appears all at once and 'fully formed'" (Gould, 1977, p. 14).

Gould was not arguing against evolution, but for his punctuated equilibrium theory. The theory avers that species maintain evolutionary equilibrium for many generations which is occasionally punctuated by rapid bursts of change. Gould does not specify a mechanism for this, but recent work has proposed that a protein called Hsp90 (heat shock protein 90) encourages stasis by acting as a chaperone for mutated genes. When Hsp90 is compromised by radical environmental change, accumulated mutations are released. Of course, only advantageous mutations are useful, and the results of one study sounded more like devolution than evolution. Researchers induced drastic temperature changes in the climate of fruit flies. After a few generations of these flies trying to adapt to the new conditions, a menagerie of insect monsters emerged: "When the genetic variations usually suppressed by Hsp90 began to express themselves, major changes developed in the insects' body plans. Some insects began to sprout weird limbs from different wings, some thick-veined wings, others deformed eyes or legs" (ABCScience, 1998).

Punctuated equilibrium is an effort to explain the conundrum of the Cambrian explosion, an event that changed the Earth's biota in profound and fundamental ways. The Earth's biota was essentially static for billions of years and then—bang! —we have: "a dynamic and awesomely complex system whose origin seems to defy explanation.... numerous animal phyla with very distinct body plans arrive on the scene in a geological blink of the eye, with little or no warning of what is to come in rocks that predate this interval of time" (Peterson, Dietrich, and McPeek, 2009, p. 736). The Cambrian conundrum was recognized by Darwin and the problem posed to his theory by the sudden appearance of numerous animal forms with no ancestors to be found in the fossil record. He wrote, "If numerous species, belonging to the same genera or families, have really started into life all at once, the fact would be fatal to the theory of descent with slow modification" (Darwin, 1982, p. 309). Not only did these body plans arrive abruptly, but their basic form has remained the same: "Evolutionary biology's deepest paradox concerns this strange discontinuity. Why haven't new animal body plans continued to crawl out of the evolutionary cauldron during the past hundreds of millions of years? Why are the ancient body plans so stable?" (Levinton, 1992, p. 84). Microevolution continues at the species level without altering basic body plans.

The "tree of life" is a metaphor Darwin used to describe the relationships between organisms, both living and extinct, to show that all life descended from a common source. Darwin admitted that the data available at the time did not support his theory, but was optimistic that future data would because he believed that the number of intermediate varieties must be "truly enormous." Having said that, he later asks: "Why then is not every geological formation and every stratum full of such intermediate links? Geology assuredly does not reveal any such finely graduated organic chain; and this, perhaps, is the most obvious and gravest objection which can be urged against my theory. The explanation lies, as I believe, in the extreme imperfection of the geological record" (Darwin, 1982, p. 292). However, 165 years later, after many thousands of paleontologists and anthropologists have spent hundreds of thousands of hours searching at hundreds of geological sites around the world, we still lack true intermediate varieties. Of course, *Homo sapiens* did not arrive full-blown at the Cambrian, but our appearance was very abrupt according to the dean of evolutionary biology, Ernst Mayr: "How can we explain this seeming saltation [the sudden appearance of new genetic characters]? Not having any fossils that can serve as missing links, we have to fall back on the time-honored method of historical science, the construction of a historical narrative" (Mayr, 2004, p. 198). The historical narrative is to take the odd tooth or bone fragment and assume them to be transitional evidence for the large, unbridged gaps in the fossil record.

Having failed to demonstrate a tree of life from the fossil record, scientists swung from bones to genes to examine nucleotide sequences from different species and to determine how closely they are related. The ultimate goal is to construct a tree of common ancestry of all species down to the roots. But despite huge amounts of data, conflicting versions of the tree are common. One suite of genes reveals one tree, while another suite reveals a quite different tree, and trees using DNA produce different results from trees using RNA. Some trees place humans in the same lineage as elephants; others in a lineage leading to worms, and other trees reveal that half of our genes have one evolutionary history and the other half a different evolutionary history (Maxmen, 2011). We also see data fudging when things don't go as planned: "because those genes produced phylogenies at odds with conventional wisdom" (Rokas and Carroll, 2006, p. 1902). Given these many difficulties, many have concluded that it has been a fool's errand. Biologist Eric Bapteste, for instance, notes: "For a long time the holy grail was to build a tree of life...A few years ago it looked as though the grail was within reach. But today the project lies in tatters, torn to pieces by an onslaught of negative evidence" (in Ebifegha, 2009, p. 36).

Theistic Evolution

Although the origin of life and macroevolution are confronted with seemingly impenetrable barriers, we cannot conclude that they will remain so. After struggling with evolutionary theory for decades, I have allied myself with theistic evolution (TE). Proponents of TE hold the belief that God created all living things using the process of evolution in ways that conform to secular scientific accounts. TE accepts both micro and macroevolution, but unlike secular accounts, it denies that they are undirected and purposeless. TE is not a scientific theory; rather, it is a view articulating how evolutionary theory relates to Christian belief and interpretation. The TE view presents a creative self-organizing universe containing laws that have made possible the existence of intelligent beings. It does not require a God who tinkers with creation, but rather a God who allows the continuous unfolding of properties invested in nature from the moment of creation. As Darwin himself noted, there is a grandeur in the idea of everything that exists emerging gradually from nothing but laws formulated by God's divine action. TE is not saying that nature acts independently of God's direction, only independently of His *direct* and immediate control. God's chosen method of bringing life into existence was the evolutionary process in which He endowed nature with the creative power to organize itself.

TE is not something that evolved in response to the challenge of Darwinism; it has been around for centuries. Augustine's *Commentary on Genesis* (V.4:11; my emphasis), written in the fourth century AD, shows that Darwinism is old news: "It is, therefore, *causally* that Scripture has said that earth brought forth the crops and trees, in the sense that *it received the power of bringing them forth*. In the earth from the beginning, in what I might call the roots of time, God created what was to be in times to come." Thomas Aquinas wrote similarly in the thirteenth century: "Nature is nothing but the plan of some art, namely a divine one, put into things themselves, by which those things move towards a concrete end: as if the man who builds up a ship could give to the pieces of wood that they could move by themselves to produce the form of the ship" (1963, p. 124). Augustine and Aquinas point out that the natural properties of the earth that make crops and trees possible (matter has been granted the power by the Creator to act on other matter without requiring His micromanagement) are secondary to the primary cause immanent in the laws of nature from the very beginning of the universe. There is little difference between them and Darwin on this; they all emphasize primary laws "impressed on matter by the Creator."

Nineteenth-century theologian Charles Kingsley also captured the idea of TE: "We knew of old that God was so wise that he could make all things: but behold, He is so much wiser than that, that he can make all things make themselves" (in Russell, 2008, pp. 167-168). And Swedish theologian Mats Wahlberg says that humans can create, but only God can create things able to create themselves: "If it takes more wisdom to create through an evolutionary process than by hands-on-design, and if structures created by hand-on-design by humans are expressive of human intent and intelligence, why could not structures created by God in that more wisdom-demanding way reflect divine intent and intelligence?" (Wahlberg, 2012, p. 182). TE is accepted by all mainstream Protestant denominations and by the Catholic Church, so it must have a strong persuasive punch.

The logic of TE is that everything evolves. Evolution simply means the gradual development of something from a simple form to a more complex form. We have no argument with the evolution of the universe. From the flash of unimaginable energy spoken into being at the Big Bang, we had the evolution of fundamental particles, which evolved into protons, neutrons, and electrons, which evolved into atoms of hydrogen and helium. These simple gases then evolved into stars and planets. Our Earth is one of those planets, but it was a barren rock before it evolved to be the complex matrix of life that it is today. TE asks that if God allowed the universe to evolve from the pure energy of the Big Bang, why not life? One may ask why God took so much time getting to the evolution of modern *homo sapiens*. God is outside of time, so what seems like an eternity to us may be just a blink of an eye for the eternal God. For TE scientists, biological evolution is a "disguised friend" of theism because it, like the discoveries of cosmology, is slowly granting us insights into how God made us.

TE is not just an "Add God and stir" approach to science. Although some TE scientists are content to accept the science and rest content with ruminating on its theological implications, others want to provide an explanation for God's guiding hand in evolution. Some have appealed to quantum mechanics, viewing God as operating at the quantum level: "The idea of a God as an agent of downward causation has emerged in quantum physics" (Goswami, 2014, p. 22). Kenneth Miller is among those who take a quantum approach: "The indeterminate nature of quantum events would allow a clever and subtle God to influence events in ways that are profound, but scientifically undetectable to us. Those events could include the appearance of mutations, the activation of individual neurons in the brain, and even the survival of individual cells and organisms affected by the chance processes of radioactive decay" (Miller, 1999, p. 241). Quantum physicist Robert Russell makes the TE case in his theory of NIODA (non-interventionist objective divine action) in which he sees continuous creation arising indirectly from "God's direct action of sustaining in existence quantum systems and their properties during both their time evolution and their irreversible interactions" (Russell, 2008, p. 590). Russell maintains that God guides creation by quantum collapse, and that small quantum effects have phenotypic consequences that will spread throughout a mating population by "biological amplification"—small micro-events amplifying into significant macro-events.

Ignacio Silva does not like the NIODA view, arguing that it reduces God's activity to a "cause-among-causes" by constraining Him to work with quantum phenomena: "God is bound by nature to act within the laws of nature" (Silva, 2015, p. 107). In other words, God is a necessary but not sufficient cause and does not act entirely autonomously. Another way of looking at NIODA is physicist John Polkinghorne's appeal to *kenosis*, which means that God willingly surrenders some of his authority and grants nature the freedom to make itself by natural forces as He grants us free will to make ourselves: "The play of life is not the performance of a predetermined script, but a self-improvisatory performance by the actors themselves.... God shares the unfolding course of creation with creatures, who have their divinely allowed, but not divinely dictated, roles to play in is fruitful becoming" (Polkinghorn, 2001, p. 94). Polkinghorn's God is thus a God who is constantly working creatively through the unfolding of the inherent potentialities in nature.

In the TE view, evolution is God's way of maintaining epistemic distance between Himself and His creation. The sudden creation of all living species in their current form would have been so obvious that it would jeopardize human autonomy and stifle biological science. TE scientists maintain that while we can see God's hand in evolution, we have to look long and hard. John Hick avers that epistemic distance is necessary for humans to *voluntarily* come to God. Hick writes, "the reality and presence of God must not be borne in upon men in the coercive way in which their natural environment forces itself upon their attention. The world must be to man, at least to some extent, *etsi deus non daretur*, 'as if there were no God.' God must be a hidden deity, veiled by his creation" (Hick, 1977, p. 281).

Whatever science eventually discovers, there is nothing that would cast doubt that only a Divine Hand could possibly be responsible for the immaterial information coded in the book of life. No one knows how He did it, but inference to the best explanation tells us that He did. The final word is left to Albertus Magnus, thirteenth-century scientist and theologian: "In studying nature we have not to inquire how God the Creator may, as He freely wills, use His creatures to work miracles and thereby show forth His power; we have rather to inquire what Nature with its immanent causes can naturally bring to pass" (in Walsh, 2013, p. 338).

Chapter Thirteen
I Am Fearfully and Wonderfully Made

The Most Complex System in the Universe is You

Psalm 139:14 says: "I praise you because I am fearfully and wonderfully made; your works are wonderful, I know that full well." Augustine echoed this when he said: "Men go abroad to wonder at the height of mountains, at the huge waves of the sea, at the long courses of the rivers, at the vast compass of the oceans, at the circular motion of the stars, and pass by themselves without wondering." We do express awe and wonder with many things without ever realizing that we are more worthy of wonder than any of them. We are stunningly engineered creations of interrelated functionality blessed with the ability to self-repair and to make others of our kind. The human body's interconnected systems are a marvel of biomechanical and biochemical engineering, converting food into energy and living tissue that repairs itself, and a million other things automatically every minute of every day. Werner Gitt notes that the human body is the most complex information-processing system in the universe and if we take the conscious information processes such as thought, speech, and deliberate voluntary movements, combined with unconscious ones (automatically controlled processes): "this involves the processing of 10^{24} bits daily. This astronomically high figure is higher by a factor of 1,000,000 than the total human knowledge of 10^{18} bits stored in all the world's libraries" (Gitt, 1996, p. 187).

Your life began when one of your father's sperm cells won the frantic race to merge with your mother's ovulated egg to form a zygote containing all the genetic information to enable you to spread from a microscopic dot to the complex creature that you are. It is a miracle that the single-cell zygote develops into an adult human being with trillions of cells because how do all these rapidly dividing cells know what to become since they all contain the same DNA and thus can potentially become any body part? The undifferentiated stem cells become differentiated by exposure to signals from both inside and outside the cells—chemicals, extracellular proteins, hormones, and neighboring cells—that turn genes on or off. Receptors on the cell surface read these signals and respond appropriately. Genes that are turned on make proteins specific to the body part they will help to fashion, and once they are differentiated the genes for other parts are forever shut down in those cells.

The master regulators of the process of building a body to the correct specification are a group of 39 genes called *Hox genes*. Hox gene expression is determined by a gene's position within a cluster and is activated by proteins encoded by earlier genes. When activated, they begin laying out the body's architecture in an orderly fashion along the embryo's head-to-tail axis. The developing body must balance a menagerie of finely-tuned kinetic, chemical, cellular, and genetic signals to make a new and distinct human being. The zygote/embryo/fetus is not developing *into* a human being; it is developing *as* the human being it already is. Just because I am further along that journey than my son, it does not make me more human than he is, nor he more than his son is. How about the human life in a young woman's womb; is it any less human than she is just because it has not yet taken its first step? The Bible states that human life is a human being from conception. Jeremiah 1:5 says: "Before I formed you in the womb I knew you, before you were born I set you apart; I appointed you as a prophet to the nations."

Sex and Zygotes

The zygote marks the beginning of human life, but its existence is one of the deepest mysteries of biology because, as mixtures of two genomes resulting from the sex act of males and females, it is incredibly improbable. Sexual reproduction is a relative newcomer on a geological time scale because reproduction was asexual for eons before it arrived. Many forms of life propagate asexually (bacteria, archaea), some both sexually and asexually (aphids, plants), but higher forms of life exclusively reproduce sexually. The mystery is why something as time-consuming and energy-depleting as sex was substituted when the more efficient asexual reproductive method had been in place for about two billion years prior to its arrival. Nature puts a high premium on genetic fidelity, so asexual reproduction should be the way to go since it transmits the entire parental DNA intact whereas sexual reproduction involves the genetic reshuffling of gametes from male and female genomes.

One theory of sexual reproduction is the "Red Queen hypothesis." It is taken from Lewis Carroll's *Through the Looking Glass* in which the Red Queen has been running with Alice for a long time and getting nowhere, stops and pants: "it takes all the running you can do to keep in the same place." The metaphor emphasizes that species must continually evolve ("keep running") to keep pace with predators and parasites. If a species stops evolving it loses the competition with species that do not. For example, gazelles either evolve genetic mutations to help them run faster than cheetahs or become extinct. If every gazelle ran faster than every cheetah, cheetahs may become extinct, so cheetahs would have to evolve greater speed. These mutations result from the constant genetic shuffling and recombination involved in sexual reproduction.

The Red Queen hypothesis describes a process that takes place *after* sexual reproduction is present, not *why* we have it in the first place. Geneticists Gorelick and Heng note that although the recombination of genes in sexual reproduction produces variety *within the species*, the primary function of sexual reproduction is to keep the species' genome intact. That is, at the species level sexual reproduction serves to maintain the species' genome while increasing variation at the individual level. Sexual reproduction is thus a brake to macroevolution while permitting microevolution, thus allowing organisms to adapt to their environments while remaining "within kind" (Gorelick and Heng, 2011). But this theory explains the advantages of sexual reproduction *once it is here*; not *how* it got here. The key to explaining the jump from asexual to sexual reproduction is to explain how meiosis evolved from mitosis, a problem that bothered W.D. Hamilton, one of the great biologists of the twentieth century. Hamilton wrote: "if there is one event in the whole evolutionary sequence at which my own mind lets my awe still overcome my instinct to analyse, and where I might concede that there may be a difficulty in seeing a Darwinian gradualism hold sway throughout almost all, it is this event—the initiation of meiosis" (Hamilton, 1999, p. 419).

The cells in our bodies are constantly being replaced by one of two forms of division: mitosis and meiosis. Mitosis produces two identical diploid cells with all the same genetic information as the parent cell. Meiosis is a more complicated process characterized by one round of DNA replication containing all 46 chromosomes. In the next round, that cell divides again, and then again, to produce four haploid cells containing half (23 chromosomes) the original amount of genetic information of the parent cell. During this process, there is a lot of genetic shuffling to ensure a new, genetically distinct, human being. Meiosis produces only gametes—female eggs and male sperm. Meiosis is alleged to have evolved from mitosis to achieve the miraculous ability to half the chromosomal count to make sexual reproduction by the union of male and female gametes possible. The union of 23 chromosomes from each of the male and female gametes provides the zygote with the full complement of 46 chromosomes.

Commenting on the strangeness of meiosis, biologist Mark Ridley points out: "You only have to think of sex to see how absurd it is. The 'sexual' method of reading a book would be to buy two [male and female] copies, rip the pages out, and make a new copy [the zygote] by combining half the pages from one and half from the other, tossing a coin at each page to decide which original to take the page from and which to throw away [random genetic shuffling]" (Ridley, 2001, p. 209). Thus, for sexual reproduction to occur, a slew of ridiculously implausible things must occur. First, males and females had to evolve; either simultaneously, or one sex evolved in order to evolve the other while still retaining the first. Then that had to evolve the internal and external sex organs and the sperm and eggs to fashion an offspring. Meiosis would have to evolve from mitosis to ensure that only 50% of each parent's genetic material is passed on. All this would have had to occur simultaneously because one sex without the other isn't much use. Additionally, the female had to evolve a mechanism to prevent her immune system from destroying the male sperm, which it would otherwise recognize as an invading antigen.

The Innate and Acquired Immune Systems

When the zygote becomes a newborn baby, it is confronted with a host of alien invaders called antigens; short for *anti*body-*gen*erating. To defend itself against these invaders, the child possesses an elaborate communication system that works around the clock to fight them off by creating antibodies specific to each antigen called the immune system. The child's first line of defense is the mother herself. While the baby was in her womb, the child developed its innate immune system, which is further primed during the birthing process when bacteria from mom's vagina are passed on to the baby, helping to build a colony of "good" bacteria in the gut. After birth, the symbiotic relationship between mother and child becomes very important.

The bliss on a mother's face as she breastfeeds the contented infant snuggled in her arms is as close to *agape*—the unconditional love of God; pure selfless love—that most humans will ever know. The look of pure delight on the mother's face is the infant's primordial experience of love—one person taking pleasure in the existence of another. The infant is experiencing warm tactile sensations snuggled in its mother's arms as it closes its lips around her nipple to drink in the milk of human kindness. The warm tactile feeling when mothers cuddle their infants sends signals to their brains' pleasure centers telling them all is right with the world. Infants can only "think" with their skin, so mothers are the fountainhead of all their satisfaction. The benefits of breastfeeding to the infant are not limited to psychology; it confers immunological benefits also. Although infants received some placental antibodies circulating in the mother's system, their immune systems are underdeveloped. Fortunately, breastfed infants receive a 100% safe vaccine in the form of colostrum, mothers' "high octane" milk. After a period of ingesting this superfood, the infant gets 10 to 14 days of transitional milk, which will eventually be replaced by regular breast milk containing many compounds important for brain development. This natural formula contains elements that support your baby's immune system such as proteins, fats, and sugars, and transfers the mother's developed antibodies to the nursing infant since mothers and their babies are typically exposed to the same antigens.

Most antigens can be handled by the innate immune system, and after birth, the infant will slowly begin to develop its more efficient adaptive immune system. The adaptive system generates a vast repertoire of defenses by producing antibodies specific to the antigens which "remember" the antigens so they can fight them more efficiently if they invade again. It is constantly on a war footing, and to identify and repel invading antigens, it has an army of spies, commandos, generals, and brigades of fierce soldiers to protect the body from harm. A key feature of the adaptive system is the ability to recognize self from non-self. Like distinctive uniforms of armies, each cell in your body carries a distinctive molecule marker that identifies it as "self" as opposed to alien cells that have invaded. Two exceptions to this are the foreign male sperm and the zygote/embryo. Because the zygote/embryo is a new individual with its cells tagged as such, they are foreign markers not recognized by its mother's cells as "self." To prevent the mother's immune response from attacking and destroying them, her immune response must be suppressed, but not completely, or else her health, as well as the baby's, would be in jeopardy. Thus, maternal immune suppression is localized at the implantation site of the uterus.

The key players in the immune system are white blood cells, or lymphocytes, made in the bone marrow. When lymphocytes are manufactured, they travel to the lymph organs to await orders to sally forth and fight for our survival. Lymphocytes come in two varieties: T-cells and B-cells. These chemical warriors can fit every shape of invading antigens. T-cells recognize invaders by the surface shape of their molecules, and in some instances, T-cells kill an invader directly, in others it sends signals to headquarters via chemical messengers called cytokines so the immune system as a whole can decide the most effective weapon(s) to use against them. These weapons include killer T-cells, B- cells, and macrophages. Macrophages ("big eaters") are giant circulating white blood cells that reach out and gobble up cells damaged by the invader and release their protein fragments that are then attacked by killer T-cells.

When B cells encounter these fragments, it manufactures an antibody protein that attaches itself to them. Each antibody matches a specific antigen in a lock-and-key fashion and marks it for destruction. Once millions of lock-specific antibodies are made, the immune system remembers the invading locks if they encounter them again and can more quickly and efficiently destroy them. All this activity is guided by another type of T-cell called a helper T-cell, which, like a field commander, coordinates the activity of all other cells, and orders the making of different types of cytokines. When the invaders have been surrounded, pounded, and shredded, there must be a mechanism to tell the good guys to cease fire lest they continue pouring troops into the fray and inadvertently attack self-cells. Suppressor T-cells cells perform this task when they perceive lots of enemies wounded and dead and release their own chemical signals to ceasefire. This is yet another functionally integrated system whose origin we struggle to explain.

The Cardiovascular System: Pipeline of Life

The cardiovascular system is the first functioning system babies develop. The blood vessels begin forming about 14 days after conception, and the heart starts pumping within 21 days. This amazing pump has four chambers, two at the bottom that pump blood carrying oxygen out of the heart, and two at the top that receive the returning blood. Returned blood gets sent to the lungs to collect more oxygen and to drop off carbon dioxide which is then exhaled. With each heartbeat, the heart sends blood around the body's pipelines (arteries that carry the blood away from the heart and veins that carry it back) which, if laid end-to-end would stretch out to about 80,000 miles. The system is so efficient that every cell in the body is serviced every 20 seconds at least 2.5 billion times in an average lifetime. Sherwin Nuland notes that "the cardiac cycle is no more than yet another of those wonders of coordination and timing upon whose flawlessness and predictability all human life depends" (Nuland, 1997, p. 213).

Blood itself is remarkable stuff, consisting of different types of cells in a protein-rich plasma fluid. The cells transported throughout the body by the plasma are red blood cells, white blood cells, and platelets. If we think of the white blood cells as the body's military, we may think of the red blood cells as the engineers, physicians, craftsmen, and pickup and delivery guys carrying out the mundane tasks of carrying life-giving supplies to cells throughout the body, collecting their wastes, and repairing problems that may arise. Unlike the white blood cells that take on all kinds of shapes, all red blood cells are shaped like doughnuts minus a central hole. Like their white brothers, red blood cells are produced in the bone marrow at the rate of about two million per second. They only last for about four months because they expel their nuclei before leaving the marrow and thus cannot clone themselves. As they age and die, they are scavenged by white blood cells and broken up. Their iron content is transported back to the bone marrow and used again to produce new red blood cells; the cellular waste is then excreted.

Finally, we have the platelets, which comprise the bulk of the blood and help the blood clotting process. Blood clotting involves a complex series of biochemical processes to ensure that it clots when and where it needs to, and when and where it doesn't. Blood starts clotting when platelets gather at the site of an injury and forms a coagulation platform. A protein called fibrin forms a crisscross scaffolding that prevents further bleeding and upon which new tissue forms. There is a tightly controlled regulatory system akin to that of the immune system so that clotting is site-specific and does not spread to healthy tissue causing heart attacks or strokes. This regulation is revealed in studies of so-called "knock out mice" in which researchers inactivate ("knockout") genes of either the "go" or "stop" phase of the process. Knock out the gene for fibrinogen, the precursor of the "go" protein fibrin, and mice hemorrhage; knock out the gene for another protein called plasminogen, the precursor of plasmin (the "stop" signal) that degrades clots, clots form throughout the circulatory system, and the mice die. These genes had to come online together for animals to survive, but how would a mindless process know that the body would need both an accelerator and a brake? Only an intelligent Being would know such a thing.

The Eyes: Windows to the World

We behold the wonders of the universe through photons that bathe photoreceptors in the eye's retina. The optic nerve moves the photons to the occipital lobe at the rear of the brain which receives, organizes, and interprets the patterns they generate. There are about 120 million photoreceptors called rods in the retina and some six or seven million cones. Computer engineer John Stevens compares the ability of our eyes to a computer, and remarks that to simulate just 10 milliseconds of the processing of a single nerve cell from the retina "would require the solution of about 500 simultaneous nonlinear differential equations 100 times and would take at least several minutes of processing time on a Cray supercomputer. Keeping in mind that there are 10 million or more such cells interacting with each other in complex ways, it would take a minimum of 100 years of Cray time to simulate what takes place in your eye many times every second" (Stevens, 1985, p. 287).

One wonders what series of mutations could explain the simultaneous origin of the optical system linked to nerves that conduct signals to the back of the brain that no super-camera or computer can match Yet some are determined to find flaws in the eye's design: "Any engineer would naturally assume that the photocells would point towards the light, with their wires leading backwards towards the brain. He would laugh at any suggestion that the photocells might point away from the light, with their wires departing on the side nearest the light" (Dawkins, 2006, p. 93). But such reasoning is in error because what appeared to be wrong wiring is exactly what is required for visual acuity. Labin and Riback tell us that the wiring once considered a source of distortion improves the resolution of the eye and reduces aberrations (Labin and Ribak, 2010, p. 4).

The Brain: The little Universe Within

The human brain is God's magnum opus; the most immensely complicated and awe-inspiring entity in the universe. The human brain is so complex that we may never understand it. As it has been said many times in neuroscience circles: "If the human brain were so simple that we could understand it, we would be so simple that we couldn't." Roger Sperry rhapsodized about the brain's complexity and mystery when he wrote that: "In the human head there are forces within forces within forces, as in no other cubic half-foot of the universe we know;" (in Fincher, 1982, p. 23) and Roger Penrose wrote that our brains are a tiny part of the cosmos: "But it is the most organized part. Compared to the complexity of the brain, a galaxy is just an inert lump" (in Holt, 2018, p. 178). The brain constitutes only 2% of body mass, but 50 to 60% of our genes are involved in its development and it consumes a voracious 20% of the body's energy resources. We could exhaust a whole dictionary of metaphors to sing the praises of this enchanted loom because within its buzzing chemical soup and electrical sparks lie our thoughts, memories, desires, emotions, intelligence, and creativity. Everything we do engages the brain as its electro/chemical circuitry captures our genetic dispositions and environmental experiences and blends them into a self-conscious human being.

The brain is a very big thing contained in a very small space (think of compressing a ton of coal into a teacup). It contains at least a trillion cells, with about 100 billion being the neurons that receive sensory input, process it, and facilitate responses to it. The foundation for this organ of unmatched complexity begins three weeks after conception when a sheet of embryonic cells called the neural plate folds and fuses into the neural tube. At his point, we can discern the four bulges that will eventually become the central nervous system: forebrain, midbrain, hindbrain, and spinal cord. The neural tube grows and cells differentiate into their assigned parts throughout the first trimester as a series of protein signals direct waves of migrating cells to their allotted place. Once embedded, neurons develop a means of communicating with other neurons by converting environmental input into electrochemical impulses within neural networks.

Each neuron has one axon, and numerous dendrites, which are branched extensions of the cell (see Figure 13.1). Axons are coated with a myelin sheath made of cholesterol which acts like insulation around electrical wiring, serving to amplify nerve impulses and to protect axons from short-circuiting. Myelin is formed from numerous glial cells which also provide the cells' physical support and nourishment. Axons transmit signals to other neurons, which are then passed on to the next neuron in the chain. Myelinated axons transmit impulses 100 times faster than unmyelinated axons. Dendrites serve as receivers of impulses from neighboring neurons. Neurons pass their information along the axon in the form of electric signals via the exchange of charged atoms (ions) in and out of the permeable membrane until they reach the axon's presynaptic knob. The message is then changed from electrical to chemical.

The chemical messengers are neurotransmitters stored in tiny vesicles that open up and spill them out into the gap (about 3,000 times smaller than the width of a human hair) between sending and receiving axons. When neurotransmitters cross the synaptic gap to make contact with postsynaptic receptor sites, the message is translated back into electrical form for further transportation or inhibition, depending upon the ratio of excitatory to inhibitory messages it receives. A neuron has the capacity to make thousands of synaptic connections with other neurons, and the brain's 100 billion neurons "form over 100 trillion connections with each other—more than all of the Internet connections in the world" (Weinberger, Elvevag and Giedd, 2005, p. 5). Neurotransmitters remain in the synaptic gap for only 1/500 of a second, with excess amounts pumped back into the presynaptic knob or degraded by enzymes to prevent signal confusion when the next signal arrives.

Figure 13.1. The Neuron and its Parts

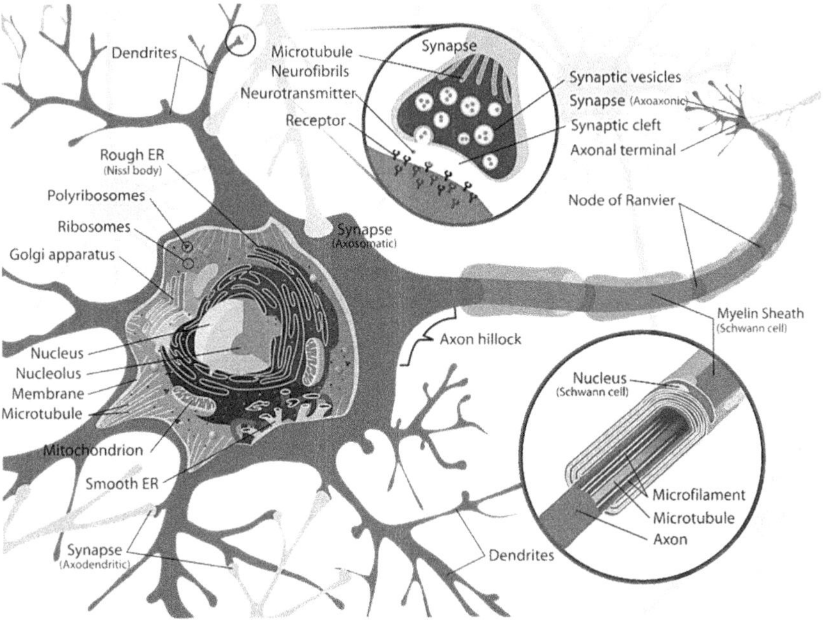

Neuroscientists have tried to follow the logic of this process using high-performance computer software and found just how difficult it is for us to duplicate it. They created a huge artificial neural network of 1.73 billion nerve cells connected by 10.4 trillion synapses, which is just a tiny fraction of the neurons and synapses the human brain contains. It is no surprise that researchers were not able to simulate the brain's activity in real time: "It took 40 minutes with the combined muscle of 82,944 processors in K computer [a supercomputer in Kobe, Japan] to get just one second of biological brain processing time. The simulation ate up about 1PB [1 petabyte (PB) is equal to 10^{15} bytes] of system memory as each synapse was modeled individually" (Scornavacchi, 2015 p. 14). Forty minutes, almost 83,000 processors, and a quadrillion bytes of memory to get just one second of the output that our billions of neurons produce every day. This shows that, although they both process information, why comparing brains to computers is wrongheaded. Miguel Nicolelis remarks that: "the brain's activity is the result of unpredictable, nonlinear interactions among its multiple billions of cells...You could have all the computer chips ever in the world and you won't create a consciousness" (in Regalado, 2013, np).

Synaptogenesis

Brain development is a process of creating, strengthening, and discarding connections among neurons called synapses. Although the wiring patterns needed for survival, such as brain areas controlling heart rate, breathing, and body temperature are present at birth, subsequent wiring is primarily a matter of experience. The process of wiring the brain by experience is known as synaptogenesis. Environmental experience *alters* synaptic patterns in adults, but it literally *organizes* the infant brain. At the peak of synapse creation, the cerebral cortex of a normal healthy infant may create up to two million synapses per second, and by two years of age, it will have approximately 100 trillion. During the first few months of an infant's life, dendrites proliferate and glial cells wrap around axons to begin the process of myelination. Dendrite growth and axon myelination continue throughout life but proceed at an explosive rate during infancy and toddlerhood.

Synapses are created and eliminated throughout life, but creation exceeds elimination in the first two years, after which production and elimination are roughly balanced until adolescence when elimination exceeds production again. In the earliest stages of synaptogenesis, dendrites send out their feelers looking for any available partners. Frequent synaptic couplings establish functional connections between neurons like the establishment of a trail in the wilderness. The more often the trail is trodden, the more distinct it becomes from its surroundings, and the easier it is to follow. Synapse retention is thus very much a use-dependent process, with only the connections that exchange information most frequently and strongly being preserved. Frequently activated neurons become primed to fire at lower stimulus thresholds once electrochemical tracks have been laid down for the impulse to follow. Experiences with strong emotional content are accompanied by especially strong electrochemical impulses and become more sensitive and responsive to similar stimuli in the future. This process is summed up in the pithy saying: "The neurons that fire together, wire together; those that don't, won't."

Humans are born with undeveloped brains relative to any other species. Infant mammals can walk within hours or days and can fend for themselves not long after. Human babies do not walk until they are about one, and cannot fend for themselves for years. The neural tube is formed from the same embryonic tissue as the skin, making the brain and skin intimately connected. The more infants are cuddled and stroked the more abundant their synaptic connections are. The expansion of two individual natures by love in this fashion; each enriching the other, was God's intention. Because the brain is the physical home of the mind, it should be primed by the power of love. Love is a powerful force; the noblest, most beautiful, and most meaningful experience of humanity. By it we are born, through it we are sustained, and for it we may sacrifice life itself. Love insulates the child, brings joy to youth, and comfort and sustenance to the aged. Its boundless power cures the sick, raises the fallen, and comforts the tormented. God created love; the magnetism between male and female; the unending devotion of parents and children, and the active concern for the well-being of others. God is love, but as an old Jewish proverb has it: "God cannot always be everywhere, and so He created mothers." Mothers are our first environment; first in her womb and then at her breast. Motherlove is not *eros*; a love that depends on the lovable qualities and virtues of the person loved, but on *agape*, a "giving" love rooted in the relationship between the lover and the loved. A mother's love for her child is like God's love for us. God's love for us is not rooted in our virtue or desert but rather in His nature as *agape* itself.

The body is indeed a wonder of design; an ecological system within which all the organs of the body are mutually dependent and harmonized so well that things go right most of the time for our allotted three score years and ten. As the great cosmologist Allan Sandage noted:

> I am convinced that the existence of life with all its order in each of its organisms is simply too well put together. Each part of a living thing depends on all its other parts to function. How does each part know? How is each part specified at conception? The more one learns of biochemistry the more unbelievable it becomes unless there is some type of organizing principle—an architect for believers—a mystery to be solved by science. (Sandage, 1985. p.54)

CHAPTER FOURTEEN
Mind, Consciousness, Language, and Free Will

I Think, Therefore I Am

Humans seem naturally inclined toward materialism because we live in a materialist world and thus tend to think in terms of space and form, and with stuff we can see, hear, taste, and touch. The immaterial realm, on the other hand, is rarely thought about, and when it is, many are inclined to dismiss it. It is thus no surprise that we get the impression from the electrochemical shunting around of molecules in the brain that they fully determine behavior in bottom-up fashion. Yet no one has the slightest idea how anything material like the brain could be conscious of itself. Materialism relegates the conscious mind to brain activity with no independent existence; "A computer made of meat," as some neuroscientists like to call it. They say this because thought cannot occur without the brain and because brain states are changed by drugs and other substances. The brain and the mind are co-dependent, but while we can have a brain without a functioning mind, we cannot have a mind without a brain any more than we can see without eyes. This may be true, but there is a large and growing literature that affirms that the conscious mind exists as an abstract entity separate from the brain.

The idea that mind and body are two distinct things with different essential qualities is an ancient notion, but the most well-known version is credited to the seventeenth-century French philosopher Rene Descartes. Descartes said that we can doubt anything except that we doubt. Doubting is thinking, so the only thing we can be absolutely certain of is that we think, as his famous dictum asserts: "*Cogito ergo sum*" ("I think, therefore I am"). For Descartes, the body is an extended material unthinking thing subject to mechanical laws, while the mind is an unextended (it transcends space) immaterial thing that thinks, and is not subject to mechanical laws. Materialists don't like this view, but David Chalmers, who has researched and written more on consciousness than any other scientist alive, argues that the conscious mind is not logically dependent entirely on the brain, and thus cannot be reduced to it. Instead, he argues that consciousness supervenes on the brain in some unknown way. He writes: "I resisted mind-body dualism for a long time, but I have now come to the point where I accept it, not just as the only tenable view but as a satisfying view in its own right... I can comfortably say that I think dualism is very likely true" (Chambers, 1996, p. 357).

The configurations of our brain states occur in top-down fashion in response to our goals, beliefs, and desires in a way not captured by bottom-up descriptions. Take the delicious feeling of romantic love. What was it about Juliette that attracted Romeo and led him to focus his desire on her? What was going on in his mind that made him love her so intensely that he would die for her, which in fact he did? We may talk of the hormone of lust (testosterone), the neurochemicals of romantic love (dopamine, serotonin, and norepinephrine), and of attachment (oxytocin and vasopressin), but these are the biochemical tools love makes use of and not the emotion that Shakespeare called "winged cupid painted blind." To say that Romeo had no choice but to fall in love with Juliette, and that love is reducible to, and determined by neurochemistry is plainly wrong. After all, he did choose her over Rosaline, and when he did his neurochemistry *responded* appropriately by translating the abstract language of his mind to the material language of his brain. All purposeful human actions evidence this pattern. Reductionist accounts of love only describe what goes on in the brain *after* people are in love, and in no way are they descriptions of *why* they are in love. The relevance of the brain activity for lovers is not mechanistic; the relevance is the information in their minds about what each represents to the other. Transcribing the mind to the brain's chemical soup and electrical sparks reduces humans to mere machines. Mental and physical events are intimately connected, but it is the conscious mind that organizes the bombardment of external stimuli that gives meaning to the brain's physical activity. The brain cannot bootstrap itself into meaningful activity by purely internal processes; it must interact with information from outside itself. Just as your car gets you from A to B without making you go there, so does your brain facilitate your mental journeys without making you take them.

Our brains allow us to function as free agents by providing the mechanisms by which we emote, think, believe, desire, plan, and worship, but it is our minds that instigate these things, and then the brain facilitates them. All human actions require a mind that forms intentions and an acting agent to carry them out. The contents of the mind form intentions and physical brain states reflect those intentions. Minds exercise causal agency; we pick up a spade to dig a hole or strike a computer key to make a letter, all at the instigation of mind, and our extended bodies are the instruments by which we accomplish such tasks. That an immaterial mind has such power is not such an abstract idea as it may seem. God is disembodied mind, and since we are created in His image, our minds are disembodied after death. The brain is a very complicated piece of biological machinery, but it cannot understand why that which it perceives gives rise to an intentional action; only a mind can do that.

Just as the lungs exist so that we may breathe, the brain exists so that we may "mind." Studies of London taxi drivers provide compelling evidence of the brain's ability to transform itself in response to mental activity. London cab drivers have to constantly learn new routes in a city of over 600 square miles with streets laid out in a spaghetti-like snarl. Using functional magnetic resonance (fMRI) machines, researchers find that cab drivers have significantly larger hippocampi (the organs of memory) than Londoners employed in other working-class occupations, and the longer they had been employed as cab drivers the larger these structures are found to be (Woollett and Maguire, 2011). If mental activity changes brain structures, perhaps it is not that the mind is "just the brain at work," but rather that the brain is "just the mind at work." It is the mind that organizes the bombardment of external stimuli that gives meaning to the brain's physical activity.

Mind and Information

It may think of mental phenomena as we think of DNA—as information. Mental information is stored in the neural structures of the brain just as biological information is stored in DNA. In packaged form, both are temporarily materially housed, but when thoughts are communicated to other parts of the brain, or outward to other minds, they are converted from their material substrate to a form of immaterial energy. Think of this as analogous to writing on a computer. When I write at my computer, what I intend to say precedes the electrical patterns engaged within the computer, and then my thoughts become physically manifested on the screen and are stored in the electrical entrails of the computer. The information content is not determined by the electrical patterns within the computer, although I need the computer (or some other physical medium) to make my thoughts manifest. My thoughts are not caused by those patterns because my though preceded them. When I decide which keys to hit to make manifest the thoughts that "come to mind," a specific sequence of electrical activity is fired up within the computer; thus, my immaterial mind has acted causally on a material object. Likewise, when I write a sentence I "fire up" a specific sequence of electrochemical activity in my brain, they are reflected in the neural correlates of my mind. If I change my mind to form a different sentence, I fire up a different sequence. This is top-down causation by which physical events in my brain are caused by my mind. Every mind state is also a brain state, but mental states are not reducible to neural properties. Nobel laureate neuroscientist and brain surgeon Sir John Eccles supports this:

The more we discover scientifically about the brain the more clearly do we distinguish between the brain events and the mental phenomena and the more wonderful do the mental phenomena become. There is a fundamental mystery in my personal existence, transcending the biological account of the development of my body and my brain. That belief, of course, is in keeping with the religious concept of the soul and with its special creation by God. (Alexander III, 2015, p. 20)

Nobel laureate physicist Eugene Wigner opines that quantum physics renders materialism logically inconsistent with the mind: "one may well wonder how materialism, the doctrine that life could be explained by sophisticated combinations of physical and chemical laws, could so long be accepted by the majority of scientists. The reason is probably that it is an emotional necessity to exalt the problem to which one wants to devote a lifetime" (Wigner, 2013, p. 177). Writing of the awesome intellect of the great mathematicians, Wigner notes: "certainly it is hard to believe that our reasoning power was brought, by the Darwinian process of natural selection, to the perfection which it seems to possess" (1990, p. 3). Stressing the complexity of the mathematics involved in quantum mechanics, he adds: "It is difficult to avoid the impression that a miracle confronts us here quite comparable in its striking nature to the miracle that the human mind can string a thousand arguments together without getting itself into contradictions or the two miracles of the existence of the laws of nature and of the human mind's capacity to divine them" (Wigner, 1990, p. 7).

Consciousness

How did we acquire a mind capable of such complex reasoning? How did this lump of mushy tissue acquire the capacity to be aware that it exists, to inject meaning into its circuitry, and the intellectual power to probe itself? Consciousness is *the* most difficult problem with which psychologists and neuroscientists have to wrestle. But why should consciousness be such a perversely difficult problem when nothing is more obvious than the fact that we are conscious? As obvious as its existence is, consciousness is so difficult to define because it is abstract and operates mysteriously. As an immaterial abstraction, it is a part of reality that cannot be reduced to physics and chemistry. This is anathema to the materialist notion that only material exists, so the best way out is to consider consciousness an illusion, as Daniel Dennett has said: "It's the brain's 'user illusion' of itself" (in Buckley, 2017, np). Nobel laureate physicist Robert Millikan, however, views consciousness as a gift of God: "The most amazing thing in all life, the greatest miracle there is, is the fact that a mind has got here at all, 'created out of the dust of the earth.' This is the Bible phrase, and science today can find no better way to describe it—a mind" (Millikan, 1927, p. 69).

Materialists offer no real alternative except to say this highest manifestation of life is another strange coincidence we must accept as a brute fact. Witness Stephen Jay Gould's statement: "Consciousness, vouchsafed only to our species in the history of life on earth, is the most god-awfully potent evolutionary invention ever developed. Although accidental and unpredictable, it has given *Homo sapiens* unprecedented power both over the history of our own species and the life of the entire contemporary biosphere" (Gould, 1997, p. ix). How could Gould's accident possibly have occurred when Darwinists tell us that evolution's sole purpose is the continued existence of life? Natural selection selects only those alleles that contribute to the twin goals of survival and reproduction, so all humans would need are the instincts that further these goals.

Humans have these instincts, so why add a conscious mind to the human repertoire when the rest of creation does quite well without it? Adding mind is an impediment to seeking reproductive success since humans generally seek it with moral rules that prevent them from seizing every mating and resource acquisition opportunity that presents itself. It is difficult to see evolutionary pressures being exerted to gift us with an impediment to biological fitness. Some neuroscientists try, believing that consciousness emerged from the increasing complexity of social life and that the need to navigate more social relationships led to a bigger brain. However, the human brain is much bigger relative to body size than is necessary to fulfill the Darwinian imperative and is energetically costly to maintain. Much brain energy is expended appreciating abstract reasoning in mathematics and science, and in the creation and appreciation of music, art, and the beauty of the world. This is a wonderful gift, but it does not afford our species any fitness advantage in raw Darwinian terms. Physicist John Polkinghorne likewise notes that: "Human powers of rational comprehension vastly exceed anything that could be simply an evolutionary necessity for survival, or plausibly construed of some sort of collateral spinoff from such a necessity" (in Frankenberry, 2008, p. 345).

Darwin himself had concerns about what his theory meant for the human mind. In an 1881 letter, Darwin wrote: "Nevertheless you have expressed my inward conviction, though far more vividly and clearly than I could have done that the Universe is not the result of chance. But then with me the horrid doubt always arises whether the convictions of man's mind, which has been developed from the mind of the lower animals, are of any value or at all trustworthy" (Darwin, 2005, p. 137). If our thoughts are nothing but molecules made by the accidental concatenation of atoms formed in an accidental universe, we may join Darwin in asking how we can believe them at all trustworthy. C.S. Lewis also wondered how such an accident would provide him with a correct account of anything, writing: "It's like expecting that the accidental shape taken by the splash when you upset a milk jug should give you a correct account of how the jug was made and why it was upset" (Lewis, 1986, p. 27).

Conscious Mind and Language: What Purpose?

If a conscious, self-aware, and intelligent, mind has no apparent Darwinian advantage, the proof being that no other species has been similarly blessed with it, it must have some other purpose. Gould was right to say that it has given us "unprecedented power both over the history of our own species and the life of the entire contemporary biosphere." This agrees with Genesis 1-28: "And God blessed them, and God said unto them, Be fruitful, and multiply, and fill the earth, and subdue it: and have dominion over the fish of the sea, and over the fowl of the air, and over every living thing that moves upon the earth."

Animals have a rudimentary form of awareness. They are aware of dangers, food sources, and mating opportunities, but they are not aware that they are aware. As purely instinctual beings, they are constrained to act as their instincts command. They are adapted to their environments, but they cannot create them. Because humans have a mind, they have escaped the captivity of mere instinctive responses to sensory perceptions. We enjoy an inner world that imposes order on the content of our sensory perceptions and think above and beyond them. Humans actively create their environments rather than merely adapting to them through their ability to think rationally and to communicate their thoughts to others. Language is a powerful externalizer of our thoughts, and the words that give them voice can powerfully influence others, for good or ill. Animals are able to communicate their intentions through sounds and gestures, but this is a far cry from the fine nuances of language. Language is so central to our lives that we tend not to spend time thinking about it, and is yet another phenomenon that seems to defy Darwinism.

The approximately 7000 languages linguists have cataloged all contain a universal grammar capable of generating the rules of any specific language. There are so many factors involved in language that we have difficulty explaining it in a manner consistent with the accumulation of beneficial mutations because the distance between humans and other primates in communicative ability is vast. Neurolinguist Elizabeth Bates suggests that this vast distance implies two options for the existence of language are possible: "either Universal Grammar was endowed to us directly by the Creator, or else our species has undergone a mutation of unprecedented magnitude, a cognitive equivalent of the Big Bang" (in Johnson and Potter, 2005, p. 87). A linguistic Big Bang would mean that language, with all the structures that make it possible, exploded on the scene instantaneously, defying the notion of the painfully slow gathering of beneficial mutations. I prefer the first option: we were endowed "directly by the Creator" with this great gift.

In 2014, a team of eight distinguished scientists from various disciplines reviewed efforts to understand language evolutionarily. They note that there are numerous ideas about the problem matched "by a poverty of evidence, with essentially no explanation of how and why our linguistic computations and representations evolved." The archaeological record and studies of animals have provided virtually no parallels to human language or any latent capacity for it, so we have no idea of what selective pressures could be involved. They also note there is little hope of connecting genes to linguistic processes, and write: "Based on the current state of evidence, we submit that the most fundamental questions about the origins and evolution of our linguistic capacity remain as mysterious as ever, with considerable uncertainty about the discovery of either relevant or conclusive evidence that can adjudicate among the many open hypotheses" (Hauser et al., 2014, p. 1).

For all the mystery involved, humans learn language effortlessly. "Learn" is perhaps not the right word. Children learn math and reading with some difficulty, but they *develop* language like they develop muscle; they breathe it in as if by osmosis. They may even acquire two or more languages simultaneously if exposed to them long enough while their brains are most plastic; often without the accent of one being transferred to the other(s). Darwinists may be perplexed about language, but the Bibles tell us that consciousness, mind, intellect, and language are gifts of God so that we may hear, speak, and understand His word: "The Lord God hath given me the tongue of the learned, that I should know how to speak a word in season to him that is weary: he wakeneth morning by morning, he wakeneth mine ear to hear as the learned" (Isaiah 50:4).

The capacity to find meaning in the universe and to experience God is impossible without the faculties of an intelligent conscious mind and language. We think about God with our minds and speak to Him with our tongues. Communication is a two-way exercise, but can we expect an all-powerful God, so far above us, to bother communicating with us? After all, He is not only our divine master, He is also our maker. He is thus immeasurably greater than us, but is He wholly "other?" Christians believe in a personal God and that we are made in His image, so to many theologians this implies at least some essential similarity with humankind. Johnson and Potter view the gift of language as necessary for our I-Thou communion with God, and express the idea that this is a characteristic God shares with us that we may come to know Him:

> God is conceived of as a communicator—whether in the form of the revealed word of God, the Ten Commandments, or his presence in religious experiences. Many theists have always believed that God created humanity in his image and that he desires ultimate communion with virtuous souls in some sort of afterlife existence. How would this be

possible unless the terms of human thought—a language of thought—have some intrinsic similarity to divine thought? And how could there be meaningful communion without some medium for communication? (Johnson and Potter, 2005, p. 92)

Free Will

The possession of free will issue has bedeviled philosophers for millennia but it is essential to Christian faith. Behaviorists insist that we don't have free will but keep pretending that we do, while existentialists insist that we do but keep pretending that we don't. Galatians 5:13 tells us we are free to choose: "You, my brothers and sisters, were called to be free. But do not use your freedom to indulge the flesh; rather, serve one another humbly in love." We have been endowed by God with the ability to make free choices, and those choices should be faithful to His commands, although He grants us the freedom to decide that they won't be. Materialism contends that free will is an illusion and that our thoughts and behavior are fully determined by our genes and our past experiences encoded in the wiring of our brains.

I define free will succinctly as the ability to choose a course of action independent of any outside influence, and determinism as the doctrine claiming that one's choice of action in any situation is the result of a sequence of outside causes channeled through our genetic makeup and prior experiences. Determinism is not to be confused with necessity. It is not a case of given X, Y *will* occur, but rather, given X, Y has a certain *probability* of occurring. Neither should we confuse free will with unpredictability. The modern version of this is based on Heisenberg's quantum mechanical principle of uncertainty, which has been used to affirm libertarian (unfettered) free will. Quantum effects are random but statistically predictable. By definition, we cannot control random events, and thus we cannot freely influence them, but how would you like it if people could not probabilistically predict your behavior? The random firing of neurons is one of the defining features of schizophrenia, and being in that sad condition is not my idea of freedom. As Sociologist Max Weber said, this is "the privilege of the insane" (in Eliaeson, 2002, p. 35).

If human action was just the result of pre-existing determinative forces, motives would translate into actions with the lubricated ease of an Audi engine, but it does not. Choices are burdensome; they must often be made against the gravitational pull of impulse, convention, habit, sloth, and the presence of alternative options. We often make bad ones, recognized only as such by hindsight, when all we can do is resolve not to make them again. That is what free will is; the ability to make both good and bad choices and to learn from them. I disavow absolute free will because we are subject to natural law.

Humans cannot have absolute free will, for only God is unencumbered by natural law, and is thus the only being that is absolutely self-determined.

Free will cannot be entirely free of causal chains. If free will is taken to mean action without a cause, all actions would be unpredictable, and chaos would reign. Everything in the universe has a cause (even quantum phenomena); we are of the universe, so what we think and do have causes. Agreeing to this does not commit us to strict determinism, since we are capable of causing our own thoughts and actions. If we are not responsible for behavior, then praise and blame alike are pointless, as are concepts of theology such as human uniqueness, sin, faith, and love. But a form of soft determinism is necessary for free will because if I did not think that what I do produces meaningful consequences, why would I do anything? All rational action, coaching, training, tutelage, and guidance are deterministic in the sense that they are designed to produce effects. We preach, discuss, and write books under the assumption that we can change the minds of others, whom we assume are free to accept or reject what we propose. I know that I am a free agent and that living according to that position is necessary, but I also know that my freedom is constrained or enabled by my temperament, upbringing, knowledge, conscience, physical and cognitive abilities and disabilities, the size of my bank account, and the constraints imposed on me by others. But without a belief in free will, our minds would be imprisoned in a deadly *Que sera sera* fatalism.

Causal talk about our behavior does not detract from our freedom and dignity. Say I know that Robert, whom I have known for a long time, has found a wallet containing a considerable sum of money and I predict that he will turn it in. I have made a prediction based on my knowledge of his moral character, but rather than insulting him I have praised him, and praiseworthiness and blameworthiness are the supporting pillars of the free will concept. If I said that although I have known him for years, he is a free agent, and therefore I do not know if he will turn it in, I would be insulting his moral character by implying that he may decide to keep the wallet.

If we do not have free will why do we all, including those who deny it, act as though we do? We acquire the notion that we are free agents, albeit with boundaries, through the everyday experience of our own volition. Surely, we could not fabricate the idea of free will without a basis for it in the reality of everyday experience. If this is not the case, why give medals to brave soldiers if they could not have done otherwise; why send criminals to prison if their actions were not their own, why praise writers, artists, musicians, or scientists if their creations were predetermined?

A famous denier of free will is Stephen Hawking. Citing "recent experiments in neuroscience," he flatly asserts that: "It is hard to imagine how free will can operate if our behavior is determined by physical law, so it seems that we are no more than biological machines and that free will is just an illusion" (Hawking and Mlodinow, 2010. p. 32). Hawkins told the materialist line asserting that when we sense we are making free choices it is simply a matter of becoming aware of what the brain has already decided what we should do, and we delude ourselves when we think that our free intentions caused those choices. Hawking's appeal to "recent experiments in neuroscience," refers to a series of experiments by Benjamin Libet showing that the brain registers decisions to make simple movements such as pressing buttons or moving a finger before subjects consciously decide to act. Libet had subjects perform simple tasks with EEG electrodes attached to their scalps while sitting in front of an oscilloscope (an instrument that displays and analyzes electronic wave signals) with a clock with a fast-circulating dot on the screen. Subjects were asked to move a finger whenever they liked, and then report the position of the dot when they became conscious of the decision to do so. Subjects' reports, coupled with oscilloscope readings, indicated that unconscious brain activity was observable as EEG signals about one-half second before they reported conscious awareness of their decision to move (Libet, Wright Jr., and Gleason, 1983). This finding was jumped on to paint humans as automatons adrift in a mindless universe, and it is a major problem for imbeciles like us who believe in free will.

Neuroscientist Iain McGilchrist says Libet's work "is only a problem if one imagines that, for me to decide something, I have to have willed it with the conscious part of my mind. Perhaps my unconscious is every bit as much 'me.' In fact, it had better be, because so little of life is conscious at all" (McGilchrist, 2009, p. 187). The subconscious areas of our minds are still us, and thus associated with our wills. After all, we do not make conscious decisions to put one foot in front of the other when we walk, but our decision to walk from one place to another is ours nonetheless. To use Libet-like experiments to argue against free will ignores the fact that we have a mechanism designed to enable us to act without conscious thought called the autonomic nervous system (ANS), the seat of the "fight or flight" response. If someone throws a rock at you there is no time for conscious deliberation; you simply duck. You don't perceive the threat and chew it over in your mind before deciding that it is a good idea to duck. Such a decision does not require conscious resources and is relegated to a lower level of awareness.

Libet's readiness potential reflects subjects' awareness that they have to perform a task and are readying their brains for it, just as police officers ready their brains for rock throwers when responding to a riot. Libet's observations reflect his subjects' anticipation of the experimental response rather than the actual conscious implementation of the movement. In other work, Libet instructed subjects to stop their intended actions once they became aware of the urge to carry them out. Subjects had a window of about 150 milliseconds after the subconscious urge to complete the task in which to consciously veto the urge, and no readiness potential was observed in this process. Libet called this abortive process "free won't." Our conscious free won't is at work every time we abort an automatic subconscious urge to light that cigarette, take another beer, or refuse an illicit relationship—it is freely taking control of our immediate desires. To interpret Libet's findings to mean that we are automatons ignores the fact that when decisions of importance are made in everyday life they are never as simple as "When will I move my finger." You don't automatically act when deciding to ask your girlfriend to marry you, buy a house, quit a job, or where to go on vacation. These decisions require conscious thought, including thought relating to the constraints that hamper your decision. This is what free will is; willing something and then consciously deciding if that's really what you should do.

Ironically, the man most often credited as the destroyer of free will was a firm believer in it. He wrote:

> My conclusion about free will, one genuinely free in the non-determined sense, is then that its existence is at least as good, if not a better, scientific option than is its denial by determinist theory... Such a view would at least allow us to proceed in a way that accepts and accommodates our own deep feeling that we do have free will. We would not need to view ourselves as machines that act in a manner completely controlled by the known physical laws (Libet, 1999, pp. 56-57).

Believing that one's lot in life is the result of someone or something else such as "society" only leads one into a dark sinkhole of victimhood and moral anarchy. A number of psychological experiments have shown that people who dismiss the idea of free will grant themselves the right to behave immorally. One such experiment involving college students found that students exposed to positive messages of determinism were significantly more likely to cheat on a variety of tasks than students exposed to positive messages of free will. The authors of one such study concluded that: "Much as thoughts of death and meaninglessness can induce existential angst that can lead to ignoble behaviors, doubting one's free will may undermine the sense of self as agent. Or, perhaps, denying free will simply provides the ultimate excuse to behave as one likes" (Vohs and Schooler, 2008, p. 54). Criminologists have long identified the fatalistic belief that one is the slave of outside influences as one of the features of people occupying the lowest rungs of society and that it is a belief that spawns hostility toward mainstream culture and promotes criminal behavior.

The Compatibilist Option

The position that Christians might adopt on the free will/determinism debate is to tie the notions together in compatibilism. Compatibilism insists that free will and determinism can peacefully coexist because one can believe in both without being logically inconsistent. Compatibilism does not deny that human events have a chain of prior events leading to them but avers that as long as we are free from external coercion, we have the freedom to decide how to respond to them. The criminal law of all civilized societies is compatibilist in that they hold criminals responsible for their actions yet leave space for a variety of mitigating factors (mental disease or defect, coercion, etc.) when sentencing them for their crimes.

Strong adherents of both free will and determinism view them as logically inconsistent positions and thus see compatibilism as evading the question. Compatibilists may counter this by appealing to Niels Bohr's principle of complementarity—the wave-particle dual nature of light. There was much initial resistance among physicists to this counterintuitive duality, but as it became more and more empirically endorsed it led to modern quantum theory. Albert Einstein and Leopold Infeld also wondered about the problem of whether light is a wave or a shower of photons and said that we cannot form a consistent description of it by choosing only one or the other. They concluded that: "It seems as though we must use sometimes the one theory and sometimes the other, while at times we may use either. We are faced with a new kind of difficulty. We have two contradictory pictures of reality; separately neither of them fully explains the phenomena of light, but together they do" (Einstein and Infeld, 1938, pp 262-263).

Human beings exhibit this same complementary duality. Substitute human action for light, and free will and determinism for waves and particles, and we can likewise conclude that neither free will nor determinism alone can explain human actions; we need both concepts to do so. Just as there is no longer any paradox in the wave-particle duality of light, there should be no paradox in viewing humans as both free agents and determined. Soft determinism gives us the only kind of free will worth having. It is a free will/free mind that follows the reasoned dictates of our natures, but it also lays on our shoulders the responsibility of owning our actions.

References

ABCScience (1998). Molecular basis for evolution. *Elsevier Science Channel*, November 27th. http://www.abc.net.au/science/articles/1998/11/27/17476.htm.

Abel, D. (2011). Is life unique? *Life*, *2*:106-134.

Aquinas, T. (1963). *Commentary on physics*. Blackwell, R., Speth, R., & Thirkel, E. (trans.). New Haven, CT: Yale University Press.

Aczel, A. (1998). *Probability 1*. San Diego: Harcourt Brace Jovanovich.

Alexander, E. (2015). Near-death experiences, The mind-body debate & the nature of reality. *Missouri Medicine*, *112*: 17-21.

Allen, R., and Lidström, S. (2016). Life, the Universe, and everything—42 fundamental questions. *Physica Scripta*, *92: 1-41*.

Andrews, E. (2017). *Is the Bible really the word of God? Is Christianity the One True Faith?* Cambridge, OH: Christian Publishing House.

Ayala, F., Cicerone, R., Clegg, M., Dalrymple, G., Dickerson, R., Gould, S., Herschbach, D., Kennedy, D., McInerney, J. & Moore, J. (1999). *Science and creationism: A view from the National Academy of Sciences*. Washington, DC: National Academy of Sciences.

Bailey, D. (2018). What are the cosmic coincidences? *Science meets religion*. http://www.sciencemeetsreligion.org/physics/cosmic.php

Balbus, S. (2014). Dynamical, biological and anthropic consequences of equal lunar and solar angular radii. *Proceedings of the Royal Society A: Mathematical, Physical and Engineering Sciences, 470:* 1-11.

Barnes, R. (2017). Tidal locking of habitable exoplanets. *Celestial Mechanics and Dynamical Astronomy, 129:* 509-536.

Barrow, J., and Tipler, F. (1986). *The Anthropic Cosmological Principle*, New York: Oxford University Press.

Bauchau, V. (2006). Emergence and reductionism: From the game of life to science of life. In Feltz, B., Crommelinck, M., and Goujon, P., pp. 29-40, *Self-organization and emergence in life sciences*. Dordrecht, The Netherlands: Springer.

Batygin, K., and Laughlin, G. (2015). Jupiter's decisive role in the inner Solar System's early evolution. *Proceedings of the National Academy of Sciences*, 112: 4214-4217.

Benner, S. (2014). Paradoxes in the origin of life. *Origins of Life and Evolution of Biospheres, 44*: 339-343.

Bernhardt, H. (2012). The RNA world hypothesis: the worst theory of the early evolution of life (except for all the others). *Biology Direct, 7: 1-10*.

Benzmüller, C., and Paleo, B. (2014). Automating Gödel's ontological proof of God's existence with higher-order automated theorem provers. In *Proceedings of the Twenty- first European Conference on Artificial Intelligence* (pp. 93-98). IOS Press.

Boeyens, J. C., and Comba, P. (2013). Chemistry by number theory. *Electronic Structure and Number Theory*, 1-24.

Bondi, H. 1952. *Cosmology*. Cambridge: Cambridge University Press.

Borwein, J., and Bailey, D. (2014). When science and philosophy collide in a 'fine-tuned' universe. Physics.Org. https://phys.org/news/2014-04-science-philosophy-collide-fine-tuned- universe.html#jCp.

Bryson, B. (2003). *A short history of nearly everything*. New York: Broadway Books.

Buckley, A. (2017). Is consciousness just an illusion? BBC News. https://www.bbc.com/news/science-environment-39482345

Camus, A. (1955). *The Myth of Sisyphus and other essays*, O'Brien, J. (trans.) New York: Knopf.

Carr, B. (2013). Lemaître's prescience: the beginning and end of the cosmos. In R. Holder and S. Mitton (eds.), *Georges Lemaître: Life, Science and Legacy* (pp. 145-172). Berlin, Heidelberg: Springer.

Carter, B. (1974). Large Number Coincidences and the Anthropic Principle in Cosmology. IAU 63, *Confrontation of Cosmological Theories with Observational Data*, 63:291–298.

Carey, N. (2015). *Junk DNA: A journey through the dark matter of the genome*. New York: Columbia University Press.

Cassé, M. (2003). *Stellar alchemy: the celestial origin of atoms*. Cambridge: Cambridge University Press.

Chambers, D. (1996). *The conscious mind: In search of a fundamental theory*. New York: Oxford University Press.

Chung, W., Wadhawan, S., Szklarczyk, R., Pond, S., and Nekrutenko, A. (2007). A first look at ARFome: Dual-coding genes in mammalian genomes. *PLoS computational biology*, 3: e91.

Cleaver, G. (2006). Before the Big Bang: String theory, God, & the origin of the universe. *Perspectives on science and religion*, June 3-7, Philadelphia, PA: Mexanexus Institute.

Clark, D. and Pazdernik, N. (2009) *Biotechnology: applying the genetic revolution*. Amsterdam: Elsevier.

Cliff, H. (2013). Could the Higgs Nobel be the end of particle physics? *Scientific American*. October 8th. https://www.scientificamerican.com/article/could-the-higgs-nobel-be-the-end-of-particle-physics/

Clinton, W. (2000). Remarks of the President. Office of the Press Secretary, the White House. https://clintonwhitehouse3.archives.gov/WH/EOP/OSTP/html/00628_2.html.

Coghlan, A. (21017). Planet Earth makes its own water from scratch deep in the mantle. *New Scientist*, January 27th. https://www.newscientist.com/article/2119475-planet-earth-makes-its-own-water-from-scratch-deep-in-the-mantle/.

Collins, F. (2006). *The Language of God: A scientist presents evidence for belief*. New York: Free Press.

Collins, F. (2007). Collins: Why this scientist believes in God. *CNN News*. http://www.cnn.com/2007/US/04/03/collins.commentary/index.html

Copithorne, W. (1971). The worlds of Wallace Pratt, *The Lamp*, 53:11-14

Corey, M. (2001). *The God hypothesis: Discovering design in our just right Goldilocks universe*. Lanham, MD: Rowman & Littlefield.

Coyne, G., and Heller, M. (2008). *A comprehensible universe: The interplay of science and theology.* New York: Springer-Verlag.

Craig, W. (2010). *On guard: Defending your faith with reason and precision.* Colorado Spring, CO: David C Cook.

Crick, F. (1994). *The astonishing hypothesis.* New York: Scribner.

Crick, F., and Orgel, L. (1973). Directed panspermia. *Icarus,*19: 341–46.

Darwin, C. (1982). *The origin of species.* London: Penguin.

Darwin, C. (1879). To John Fordyce, 7 May 1879. Cambridge University Darwin Correspondence Project. ghttps://www.darwinproject.ac.uk/letter/DCP-LETT-12041.xml

Darwin, C. (1892). *Charles Darwin: His life told in an autobiographical chapter, and in a selected series of his published letters.* Edited by F. Darwin. London: John Murray.

Davies, P. (1982). *The accidental universe.* Cambridge: Cambridge University Press.

Davies, P. (1983). *God and the New Physics.* New York: Penguin.

Davies, P. (1984). *Superforce: The search for a grand unified theory of nature.* New York: Simon & Schuster.

Davies, P. (2003). *The Origin of Life.* London: Penguin Books.

Davies, P. (2007). *Cosmic jackpot: Why our universe is just right for life.* New York: Houghton Mifflin Harcourt.

Dawkins, R. (2006). *The God delusion.* New York: Houghton Mifflin.

Dean-Lindsey, J. (2021). Design arguments for the existence of God. *Litteratus,* 1-4, April 15.

De Duve, C. (1995). *Vital Dust: Life as a cosmic imperative.* New York: Basic Books.

Dembski, W. (2004). *The design revolution: Answering the toughest questions about intelligent design.* Westmont, IL: InterVarsity Press.

Denton, M., Marshall, C., and Legge, M. (2002). The protein folds as platonic forms: new support for the pre-Darwinian conception of evolution by natural law. *Journal of Theoretical Biology, 219*: 325-342.

Dingle, H. (1972) *Science at the crossroads.* London: Martin Brian & O'Keefe.

Duck, M., and Duck, E. (2014). *Waters of creativity: Navigating the straits between science and theology to find the source of one's beginning.* Lake Mary, FL: Charisma Media.

Dyson, F. (1979). *Disturbing the universe.* New York: Harper and Row.

Dyson, L., Kleban, M., and Susskind, L. (2002). Disturbing implications of a cosmological constant. *Journal of High Energy Physics,* 10: 1-26.

Ebifegha, M. (2009). *The Darwinian delusion: The scientific myth of evolutionism.* Bloomington, IN: AuthorHouse.

Einstein, A. (1923). *Sidelights on Relativity (Geometry and Experience).* New York: P. Dutton.

Einstein, A., and L. Infeld (1938). *The evolution of physics.* Cambridge: Cambridge University Press.

Eliaeson, S. (2002). *Max* Weber's methodologies: Interpretation and critique. Malden, M: Blackwell.

Ellis, G. (2011). The untestable multiverse. *Nature,* 469:294-295.

Ellis, G., and Silk, J. (2014). Scientific method: Defend the integrity of physics. *Nature, 516:* 321-323.

Feulner, G. (2012). The faint young Sun problem. *Review of Geophysics,* 50:1-29.

Folger, T. (2008). Science's alternative to an intelligent creator: The multiverse theory. *Discover Magazine,* December 10th. http://discovermagazine.com/2008/dec/10-sciences-alternative-to-an-intelligent-creator.

Fowler, D., Coyle, M., Skiba, U., Sutton, M., Cape, J., Reis, S., Sheppard, L., Jenkins, A., Grizzetti, B., Galloway, J. and Vitousek, P. (2013). The global nitrogen cycle in the twenty-first century. *Philosophical Transactions of the Royal Society B: Biological Sciences, 368:*.20130164.

Frankenberry, N. (2008). *The faith of scientists: In their own words.* Princeton, NJ: Princeton University Press.

Galison, P., Holton, G. and Schweber, S. (2008). *Einstein for the 21st century: His legacy in science, art, and modern culture.* Princeton, NJ: Princeton University Press.

Garay, A. (1993). Theoretical and experimental studies of the possibility of chirality dependent time direction in molecules. In *Chemical Evolution: Origin of Life* (Ed. Ponnamperuma, C. & Chela-Flores, J.), pp. 165–179. Hampton, VA: Deepak Publishing.

Gauger, A., and Axe, D. (2011). The evolutionary accessibility of new enzymes functions: A case study from the biotin pathway. *Bio-Complexity, 2011: 1-17.*

Gefter, A. (2008). Why it's not as simple as God vs the multiverse. *New Scientist, 2685*(04).

Gingerich, O. (2014). *God's Planet.* Cambridge, MA: Harvard University Press.

Gitt, W., Compton, B. and Fernandez, J., (2011). *Without Excuse.* Atlanta, GA: Creation Book Publishers.

Gonzales, G., Brownlee, D. and Ward, P. (2001). Refugees for life in a hostile universe. *Scientific American,* 285: 60-67.

Gonzales, G., and Richards, J. (2004). *The privileged planet: how our place in the cosmos is designed for discovery.* New York: Regnery Publishing.

Gonzalez, G. and Ross, H. (2000). Home alone in the universe, *First things,* May 1. http://www.firstthings.com/ftissues/ft0005/opinion/gonzalez.html.

Goswami, A. (2014). *Creative evolution: A physicist's resolution between Darwinism and intelligent design.* Wheaton, IL: Quest Books.

Gould, S. (1997). Foreword: The positive power of skepticism, *Why people believe weird things,* Michael Shermer. New York: W.H. Freeman.

Gould, S. (1998). *Leonardo's mountain of clams and the diet of worms.* London: Jonathan Cape.

Gribbin, J. (2018). Alone in the Milky Way. *Scientific American,* 319: 94-99.

Gribbin, J. and Rees, M. (1989). *Cosmic coincidences: Dark matter, mankind, and anthropic cosmology.* New York: Bantam Books.

Grossman, L. (2011). Water's quantum weirdness makes life possible. *New Scientist,* October 25th.

Gu, L., Baldocchi, D., Wofsy, S., Munger, J., Michalsky, J., Urbanski, S., and Boden, T. (2003). Response of a deciduous forest to the Mount Pinatubo eruption: Enhanced photosynthesis. *Science,* 299: 2035-2038.

Gust, D., Moore, T., and Moore, A. (2009). Solar fuels via artificial photosynthesis. *Accounts of chemical research, 42*:1890-1898.

Hall, S. (2012). Hidden treasures in junk DNA. *Scientific American*, October 1. https://www.scientificamerican.com/article/hidden-treasures-in-junk-dna/.

Hartsfield, T. (2016). String theory has failed as a scientific theory, *Real Clear Science*, January 8. http://www.realclearscience.com/blog/2016/01/string_theory_has_failed_as_a_scientific_theory.html.

Hauser, M., Yang, C., Berwick, R., Tattersall, I., Ryan, M., Watumull, J., Chomsky, N. and Lewontin, R. (2014). The mystery of language evolution. *Frontiers in Psychology, 5*:1-12.

Hawking, S. (n.d.). The beginning of time. Stephen Hawking website. http://www.hawking.org.uk/the-beginning-of-time.html.

Hawking, S. (1988). *A brief history of time*. New York: Bantam Books.

Hawking, S. and Mlodinow, L. (2010). The *Grand Design*. New York: Bantam Books.

Hawking, S. (2001). *The universe in a nutshell*. New York: Bantam books.

Heile, F. (2016). Is it theoretically possible to build a collider that can test the predictions of string theory? https://www.quora.com/Is-it-theoretically-possible-to-build-a-collider-that-can-test-the-predictions-of-string-theory.

Hick, J. (1963). *Philosophy of religion*. Englewood Cliffs, NJ: Prentice-Hall.

Hick, J. (1977). *Evil and the God of love*. New York: Harper & Row.

Hoffman, N. (2001). The Moon and plate tectonics: Why we are alone. *Space Daily*. http://www.spacedaily.com/news/life-01x1.html.

Holder, R. (2013). Lemaître and Hoyle: Contrasting characters in science and religion. In Holder, R. & Mitton, S. (Eds.), *Georges Lemaître: Life, science and legacy*, pp. 39-54.

Holt, J. (1997). Science resurrects God. *Wall Street Journal*, December 24.

Hoyle, F. (1982). The universe: Past and present reflections. *Annual Review of Astronomy and Astrophysics, 20:* 1-36.

Hoyle, F. (1999). *Mathematics of evolution*. Memphis, TN: Acorn Enterprises.

Hoyle, F., and Wickramasinghe, C. (1981). *Evolution from space*. London: JM Dent.

Huchingson, J. (2005). *Religion and the natural sciences: The range of engagement*. Eugene, OR: Wipf and Stock.

IBM (1999). IBM announces $100 million research initiative to build world's fastest supercomputer. Press Release, December 6th. https://www-03.ibm.com/press/us/en/pressrelease/1950.wss

Ilić, I., Stefanović, M., and Sadiković, D. (2018). Mathematical determination in nature: The golden ratio. *Acta Medica Medianae, 57:* 124-129.

Innanen, K., Mikkola, S., and Wiegert, P. (1998). The Earth-Moon system and the dynamical stability of the inner solar system. *The Astronomical Journal, 116:* 2055-2057.

Isaacson, W. (2007). *Einstein: His life and universe*. New York: Simon and Shuster.

Jastrow, R. (1981). *The enchanted loom: Mind in the universe*. New York: Simon & Schuster.

Jastrow, R. (1992). *God and the Astronomers*. New York: WW Norton.

Jeans, J. (1930). *The mysterious universe*. Cambridge: Cambridge University Press.

Jenkins, A., and Perez, G. (2010). Looking for life in the multiverse. *Scientific American, 302*:42-51.

Jennings, B. (2015). *In defense of scientism: An insider's view of science.* Vancouver, BC: Byron K. Jennings.

Johnson, J., and Potter, J. (2005). The argument from language and the existence of God. *The Journal of religion, 85*: 83-93.

Johnson, P. (1999). The Church of Darwin. *Wall Street Journal*, August 16th.

Kahane, G. (2014). Our cosmic insignificance. *Noûs, 48:* 745-772.

Kainz, H. P. (2010). *The existence of God and the faith-instinct.* Selinsgove, PA: Susquehanna University Press.

Kennedy, D. (1907). St. Albertus Magnus. In *The Catholic Encyclopedia.* New York: Robert Appleton Company. New Advent: http://www.newadvent.org/cathen/01264a.htm. p. 265.

Kenyon, D. (2002). *Unlocking the mystery of life*: Script draft of video. http://www.divinerevelations.info/documents/intelligent_design/unlockingthemysteryoflif escript.pdf.

Keyser, C. (1915). *The new infinite and the old theology.* Yale University Press.

Klyce, B. (nd). The RNA world and other origin-of-life theories. https://www.panspermia.org/rnaworld.htm.

Kohler T., Pratt J., Debarbieux B., Balsiger J., Rudaz G., and Maselli D., (eds) (2012). *Sustainable mountain development, green economy and institutions. From Rio 1992 to Rio 2012 and beyond.* Swiss Agency for Development and Cooperation (SDC); Centre for Development and Environment (CDE), University of Bern.

Koonin, E. (2007). The cosmological model of eternal inflation and the transition from chance to biological evolution in the history of life. *Biology Direct, 2*: 1-21.

Korthof, G. (2006). Fred Hoyle's *The Intelligent Universe*: A summary & review. http://wasdarwinwrong.com/kortho47.htm.

Lane, N., Allen, J., and Martin, W. (2010). How did LUCA make a living? Chemiosmosis in the origin of life. *BioEssays, 32*: 271-280.

Laughlin, R. (2005). *A different universe: Reinventing physics from the bottom down.* New York: Basic Books.

Lennox, J. (2009). *God's Undertaker: Has Science Buried God?* Oxford: Lion.

Lennox, J. (2011). *God and Stephen Hawking: Whose Design is it Anyway?* Oxford: Lion Books.

Lennox, J. (2012). Not the God of the gaps, but the whole show. The Cristian Post, August 20th. https://www.christianpost.com/news/the-god-particle-not-the-god-of-the-gaps-but-the-whole-show.html

Leslie, J. (1989). *Universes.* London: Routledge.

Levinton, J. (1992). The Big Bang of animal evolution. *Scientific American,* November.

Lewis, C. (1986). *The grand miracle: And other selected essays on theology and ethics from God in the dock.* New York: Ballantine Books.

Lewontin, R. (1997). Billions and billions of demons. *New York Review of Books,* January 9th.

Libet, B. (1999). Do we have free will? *Journal of Consciousness Studies, 6*: 47-57.

Libet, B., Wright Jr, E. W., and Gleason, C. A. (1983). Preparation-or intention-to-act, in relation to pre-event potentials recorded at the vertex. *Electroencephalography and clinical Neurophysiology, 56*:367-372.

Lightman, A. (2011). The accidental universe: Science's crisis of faith. *Harper's Magazine*, December.

Lim, R. (2017). *Self and the Phenomenon of Life: A Biologist Examines Life from Molecules to Humanity*. Hackensack, NJ: World Scientific.

Lipton, P. (2000). Inference to the best explanation. In W. Newton-Smith (ed.), *A companion to the philosophy of science*. pp. 184.193. Hoboken, NJ: Blackwell.

Livio, M. (2003). *The Golden Ratio: The story of phi, the world's most astonishing number*. New York: Broadway Books.

Livio, M., and Rees, M. (2005). Anthropic reasoning. *Science, 309:* 1022-1023.

Lohr, S. (1999) I.B.M. plans a supercomputer that works at the speed of life. *New York Times*, December 6. https://www.nytimes.com/1999/12/06/business/ibm-plans-a-supercomputer-that-works-at-the-speed-of-life.html.

Maddox, J. (1989). Down with the big bang. *Nature, 34:* 425.

Marsh, J. (2012). *The Liberal Delusion: The Roots of Our Current Moral Crisis*. Bury St. Edmunds, England: Arena books.

Maxmen, A. (2011). Evolution: A can of worms. *Nature News, 470*:161-162.

McGilchrist, I. (2009). *The master and his emissary: The divided brain and the making of the western world*. Yale University Press.

McGrath, A. (2010). *Mere Theology*. London: SPCK Publishers

McIntosh, A. (2009). Information and entropy–top-down or bottom-up development in living systems? *International Journal of Design & Nature and Ecodynamics, 4:* 351-385.

McLean, E. (2017). Reasons to panic about the hierarchy problem. https://massgap.wordpress.com/2017/03/26/reasons-to-panic-about-the-hierarchy-problem/

Meyer, S. (2009). *Signature in the Cell: DNA and the Evidence for Intelligent Design*. Grand Rapids, MI: Zondervan.

Miller, K. (1999). *Finding Darwin's God*. New York: Harper-Collins.

Millikan, R. (1927). *Evolution in science and religion*. New Haven, CT: Yale University Press.

Moore, W. (2015). *Schrodinger: Life and thought*. Cambridge: Cambridge University Press.

Müller, G. (2017). Why an extended evolutionary synthesis is necessary. *Interface focus, 7,* 20170015.

Mayr, E. (2001). *What evolution is*. New York: Basic Books.

Nassour, R., Ayash, A., and Al-Tameemi, K. (2020). Anthocyanin pigments: Structure and biological importance. *Journal of Chemical and Pharmaceutical Sciences, 13:* 45-57.

National Aeronautics and Space Administration (nd). *Tests of Big Bang: The CMB*. https://wmap.gsfc.nasa.gov/universe/bb_tests_cmb.html

National Aeronautics and Space Administration (2019). Wilkinson Microwave Anisotropy Probe https://map.gsfc.nasa.gov/universe/WMAP_Universe.pdf

National Aeronautics and Space Administration (2020). *Exoplanets: The Search for life*. https://exoplanets.nasa.gov/search-for-life/habitable-zone/

National Institute of Heath (2012). ENCODE data describes function of human genome. National Human Genome Research Institute https://www.genome.gov/27549810/2012-release-encode-data-describes-function-of-human-genome/

Neveu, M., Kim, H., and Benner, S. (2013). The "strong" RNA world hypothesis: Fifty years old. *Astrobiology, 13:* 391-403.

Nitardy, C. (2012). *Stumbling blocks of evolutions.* Maitland, FL: Xulaon.

Nuland, S. (1997). *The wisdom of the body.* New York: Alfred A. Knopf.

Ofulla, A. (2013). *The secrets of hidden knowledge: how understanding things in the physical realm nurtures life.* Bloomington, IN: Abbott Press.

Olsen, B. (2013). *Future Esoteric: The Unseen Realms.* San Francisco: CCC Publishing.

Overman, D. (2008). *A case for the existence of God.* Lanham, MD: Rowman & Littlefield.

Ozturk, S., Yalta, K., and Yetkin, E. (2016). Golden ratio: A subtle regulator in our body and cardiovascular system? *International journal of cardiology, 223:* 143-145.

Page, D. (2007). Predictions and test of multiverse theories. in B. Carr (ed.), *Universe or multiverse,* pp. 411-430. Cambridge: Cambridge University Press.

Pagels, H. (1985). A cozy cosmology. *The Sciences* 25:34-38.

Penrose, R. (2010). Scientist debunks Hawking's 'no God needed' theory. *Independent Catholic* September 29. http://www.indcatholicnews.com/news.php?viewStory=16815.

Penrose, R. (2016). *The emperor's new mind: Concerning computers, minds, and the laws of physics.* New York Oxford University Press.

Persaud, C. (2007). *Evolution: Beyond the realm of real science.* Maitland, FL: Xulon Press.

Peterson, K., Dietrich, M., and McPeek, M. (2009). MicroRNAs and metazoan macroevolution: insights into canalization, complexity, and the Cambrian explosion. *Bioessays, 31*:736-747.

Planinić, J. (2010). The design argument – Anthropic principle. *Journal of Philosophy and Religious studies,* 65: 47-54.

Planck, M. (1949). *Scientific Autobiography and Other Papers* (trans. F. Gaynor). New York: Philosophical Library.

Plaxco, K., & Gross, M. (2006). *Astrobiology: a brief introduction.* Baltimore, MD: Johns Hopkins University Press.

Polkinghorn, J. (2001). Kenotic creation and Divine action, in Polkinghorne, J. (ed.). *The work of love: Creation as kenosis,* pp. 90-106. Grand Rapids, MI: Eerdmans.

Pross, A. (2012). *What is Life? How chemistry becomes biology.* Oxford: Oxford University Press.

Radford, T. (2010). The Grand Design: New answers to the ultimate questions of life by Stephen Hawking and Leonard Mlodinow, September 17th. https://www.theguardian.com/books/2010/sep/18/questions-life-cosmology-stephen-hawking

Reeves, M., Gauger, A., and Axe, D. (2014). Enzyme families--shared evolutionary history or shared design? A study of the GABA-Aminotransferase family. *BIO-Complexity,* 2014:1-16.

Regalado, A. (2013). The brain is not computable. *MIT Technology Review*, February 18th.

Robertson, M., & Joyce, G. (2012). The origins of the RNA world. *Cold Spring Harbor perspectives in biology*, 4(5), a003608.

Rokas, A., and Carroll, S. (2006). Bushes in the tree of life. *PLoS biology, 4:1899-1904*.

Ross, H. (1993). The Creator and the Cosmos: How the greatest scientific discoveries of the century reveal God. *Colorado Springs, CO: NavPress*.

Ross, H. (1994). Astronomical evidences for a personal, transcendent God. In J. Moreland (ed.) *The Creation Hypothesis: Scientific Evidence for an Intelligent Designer*, pp. 141-172. Downers Grove, IL: InterVarsity Press.

Ross, H. (2019). Solar and Lunar tides designed for complex life. Reasons to Believe, March 25. https://reasons.org/explore/blogs/todays-new-reason-to-believe/solar-and-lunar-tides-designed-for-complex-life

Russell, B. (1919). *The Study of Mathematics. Mysticism and Logic: And Other Essays*. London: Longman.

Russell, B. (1935). *Religion and science*. New York: Oxford University Press.

Russell, R. (2008a). Quantum physics and the theology of non-interventionist objective divine action. In P. Clayton & Z. Simpson (eds.), *The Oxford handbook of religion and science*, pp. 579-595. Oxford: Oxford University Press.

Russell, R. (2008b). *Cosmology from Alpha to Omega: The creative mutual interaction of theology and science*. Minneapolis: Fortress.

Sandage, A. (1985). A scientist reflects on religious belief. *Truth: An International, Interdisciplinary Journal of Christian Thought, 1:* 53-54.

Sandage, A. (1990). Interview with William Durbin.

Sanford, J., Brewer, W., Smith, F., & Baumgardner, J. (2015). The waiting time problem in a model hominin population. *Theoretical Biology and Medical Modelling, 12: 1-28*.

Schaefer, H. (2003). *Science and Christianity: Conflict or coherence?* Watkinsville, GA: The Apollos Trust.

Schafer, L. (2006). Quantum reality and the consciousness of the universe. *Zygon*, 41:505-532.

Scornavacchi, M. (2015). *Superintelligence, Humans, and War*. Joint Forces Staff College Joint Advanced Warfighting School Staff College, Norfolk, VA.

Seckbach, J., & Gordon, R. (2009). *Divine action and natural selection: science, faith and evolution*. Hackensack, NJ: World Scientific.

Shalev, B. (2003). *Religion of Nobel Prize winners. 100 years of Nobel prizes*. New Delhi: Atlantic Publishers & Distributors.

Shaviv, G. (2015). Who discovered the Hoyle Level? *Acta Polytechnica CTU Proceedings, 2:* 311-320.

Shostak, S. (2011). Who or what built the universe? *HuffPost*, May 25th.

Silva, I. (2015). A cause among causes? God acting in the natural world. *European Journal for Philosophy of Religion* 7: 99-114.

Sinha, S. (2017). The Fibonacci numbers and its amazing applications. *International Journal of Engineering Science Invention, 6:* 7-14.

Smith, W. (1981). *Therefore Stand*. New Canaan, CT: Keats Publishing.

Spradley, J. (2010). Ten lunar legacies: Importance of the Moon for life on Earth. *Perspectives on Science & Christian Faith, 62*:267-275.

Stark, R (2003). *For the Glory of God.* Princeton, NJ: Princeton University Press.

Strauss, M. (2017). The God Particle... and God. https://www.michaelgstrauss.com/2017/01/the-god-particleand-god.html.

Strobel, L. (2004). *The case for a Creator: A journalist investigates scientific evidence that points toward God.* Zondervan.

Susskind, L. (2005). *The Cosmic Landscape: String Theory and the Illusion of Intelligent Design,* New York: Little, Brown, and Company.

Świeżyński, A. (2016). Where/when/how did life begin? A philosophical key for systematizing theories on the origin of life. *International Journal of Astrobiology, 15:* 291-299.

Swindell, R. (2003). Shining light on the evolution of photosynthesis. *Journal of Creation,* 17:74-84.

Taylor, S. (1998). On the difficulties of making earth-like planets. *Meteoritics and Planetary Science, 34: 317-329.*

Tegmark, M. (2009). The multiverse hierarchy. *arXiv preprint arXiv:0905.1283.*

Tegmark, M. (2014). Is the Universe made of math? *Scientific American,* December. https://www.scientificamerican.com/article/is-the-universe-made-of-math-excerpt/

Thapa, G., & Thapa, R. (2018). The relation of Golden Ratio, mathematics and aesthetics. *Journal of the Institute of Engineering, 14:* 188-199.

Tipler, F. (1988). The anthropic principle: a primer for philosophers. In *PSA: Proceedings of the Biennial Meeting of the Philosophy of Science Association* Vol. 1988: 27-48.

Tipler, F. (1994). *The physics of immortality: Modern cosmology, God, and the resurrection of the dead.* New York: Anchor.

Trefil, J., and Hazen, R. (2007). *The sciences: An integrated approach.* New York, Wiley.

University of California, Davis (nd). The electromagnetic spectrum. Online educational course. http://earthguide.ucsd.edu/virtualmuseum/ita/07_1.shtml

Valencia, D., O'Connell, R., and Sasselov, D. (2007). Inevitability of plate tectonics on super-earths. *The Astrophysical Journal Letters, 670:*45-48.

Varghese, R. (2013). *The Missing Link: A Symposium on Darwin's Framework for a Creation evolution Solution.* Lanham, MD: Rowman & Littlefield.

Vasas, V., Szathmáry, E., and Santos, M. (2010). Lack of evolvability in self-sustaining autocatalytic networks constraints metabolism-first scenarios for the origin of life. *Proceedings of the National Academy of Sciences, 107:* 1470-1475.

Vieru, T. (2011). Moons like our own are extremely rare in the universe. Softpedia News, June 29th. https://news.softpedia.com/news/Moons-Like-Our-Own-Are-Extremely-Rare-in-the-Universe-214242.shtml.

Vohs, K., and Schooler, J. (2008). The value of believing in free will: Encouraging a belief in determinism increases cheating. *Psychological Science, 19:* 49-54.

Wald, G. (1954). The origin of life, *Scientific American,* 191: 45–53.

Wald, G. (1984). Life and Mind in the Universe. *International Journal of Quantum Chemistry, 26*: 1-15.

Wahlberg, M. (2012). *Reshaping Natural Theology: Seeing Nature as Creation.* London: Palgrave Macmillan.

Walker, S., & Davies, P. (2013). The algorithmic origins of life. *Journal of the Royal Society Interface, 10*(79), 20120869.

Walker, S., & Davies, P. (2016). The "hard problem" of life. *arXiv preprint arXiv: 1606.07184.*

Wallace, P. (2016). *Stars beneath us: Finding God in the evolving cosmos.* Minneapolis, MN: Fortress Press.

Walsh, A. (2020). *God, Science, and Society: The Origin of the Universe, Intelligent Life, and Free Societies.* Wilmington: DE: Vernon Press.

Walsh, J. (2013). *Old time makers of medicine.* New York: Simon and Schuster.

Watson, B. (2011). Setting the stage for life: Scientists make key discovery about the atmosphere of early Earth. Rensselaer Polytechnic Institute. www.sciencedaily.com/releases/2011/11/111130141855.htm

Wei-Haas, M. (2018). Volcanoes, explained. *National Geographic,* January 15th. https://www.nationalgeographic.com/environment/natural-disasters/volcanoes/

Weinberg, S. (1987). Anthropic bound on the cosmological constant. *Physical Review Letters, 59:* 2607-2610.

Weitnauer, C. (2013). The irony of atheism. In T. Gilson & C. Weitnauer (eds.) *True reason,* pp. 25-36. Grand Rapids, MI: Kregel.

Wells, J. (2017). *Zombie science: More icons of evolution.* Seattle, WA: Discovery Institute.

Wigner, E. (1990). The unreasonable effectiveness of mathematics in the natural sciences. *Mathematics and Science* 13:1-14.

Wigner, E. (2013). *The collected works of Eugene Paul Wigner: Historical, philosophical, and socio-political papers. Historical and Biographical Reflections and Syntheses.* Berlin: Springer-Verlag.

Williams, G. (1992). *Natural selection: Domains, levels and challenges.* New York: Oxford University Press.

Wood, B. (2002). Who are we? *New Scientist,* 44-47.

Woollett, K., and Maguire, E. (2011). Acquiring "the Knowledge" of London's layout drives structural brain changes. *Current biology, 21:* 2109-2114.

Yahya, H. (1999). *The Creation of the Universe.* Istanbul:Global Yayincilik.

Yalta, K., Ozturk, S., and Yetkin, E. (2016). Golden Ratio and the heart: A review of divine aesthetics. *International journal of cardiology, 214:* 107-112.

Yockey, H. (2005). *Information theory, evolution, and the origin of life.* Cambridge: Cambridge University Press.

Zalasiewicz, J. & Williams, M. (2014). Weird wet worlds: Why Earth is lucky to have oceans. *New Scientist,* October 29th. https://www.newscientist.com/article/mg22429930-600-weird-wet-worlds-why-earth-is-lucky-to-have-oceans/

Figures

Figure 3.1. The Golden Spiral. Public Domain. https://commons.wikimedia.org/wiki/File:GoldenSpiralLogarithmic_color_in.gif

Figure 4.1. Standard Model of Elementary Particles. Public Domain https://commons.wikimedia.org/wiki/File:Standard_Model_of_Elementary_Particles.svg

Figure 5.1. Expansion of the Universe. Public Domain. https://commons.wikimedia.org/wiki/Commons:Featured_picture_candidates/File:CMB_Timeline300_no_WMAP.jpg.

Figure 6.1. Electromagnetic Spectrum. NASA Public Domain. https://imagine.gsfc.nasa.gov/science/toolbox/emspectrum1.html

Figure 7.1. Habitable Zones around Stars. NASA Public Domain. https://exoplanets.nasa.gov/search-for-life/habitable-zone/

Figure 8.1. The Triple Alpha Process. Public Domain. https://commons.wikimedia.org/wiki/File:Triple-Alpha_Process.svg.

Figure 11.1. The Making of a Protein. Public Domain. U.S. Department of Energy. http//:www.ornl.gov/hgmis.

Figure 13.1. Neuron and its Parts. Public Domain. https://en.wikipedia.org/wiki/Neuron#/media/File:Complete_neuron_cell_diagram_en.svg

Index

A

A Brief History of Time 23
abduction 7
Abel, David 98
abiogenesis 91, 96
abortive process 147
abstraction 75, 99
adaptive immune system 129
adenine (A) 104-105
adenosine triphosphate (ATP) 66, 111
aerosols 67
afferent nerves 105
agape 128, 135
alleles 104, 115
amino acids 92-94
    sequencing 105-106, 110
ammonia 90, 92
anisotropy 48
Annual Review of Astronomy and Astrophysics 86
anthocyanins (ACNs) 89
anthropic principle 3-5
    Davis on 71
    Susskind on 50
antibodies 128-129
antimatter 45
apoptosis 109
Aquinas, Thomas 122
Aristotle 3
Arkani-Hamed, Nima 4
asphalt problem 95
atheism 17
    and Darwin 114
    and Kenyon 92
atmospheric nitrogen 90
atomic weight 81
atoms 31-32
    and water 81
    primeval 41
Augustine 21, 122, 125
autonomic nervous system (ANS) 146
awareness 142, 146
axon 132-134

B

Bacon, Roger 14
Balbus, Steven 58
Bapteste, Eric 121
Bates, Elizabeth 142
B-cells 129
Benner, Steven 95
Bernhardt, Harold 97
beryllium 85
beta decay 38
Big Bang 41-42
    and evolution 122-123
    and multiverse 72
    linguistic 142
    opposition 43-45
Big Bang Theory 2, 14
big whack 68
biological amplification 123
Birney, Ewan 108
blood clotting 130
Blue Gene 107
Bohr, Niels 148
Bonaparte, Napoleon 12
Bondi, Hermann 2
Boscovich, Roger 14
bosons 32
brain 131-133
    and consciousness 137-138
    and synaptogenesis 134

breastfeeding 128
Brown, Arthur 64
Bryson, Bill 109

C

Calvin, Melvin 15
Cambrian explosion 119-120
Camus, Albert 1
capillary action 82
carbon 84-86
    and photosynthesis 87-89
carbon dioxide 67, 87-88
cardiovascular system 130
Carey, Nessa 108
Carnoy, Jean-Baptiste 14
Carr, Bernard 79
Carroll, Lewis 126
Carter, Brandon 3
cause 43
cell membrane 110
cells 109-111
    B-cells 129
    diploid 127
    glial 132, 134
    haploid 127
    red blood 130
    T-cells 129
    white blood 130
Chalmers, David 137
chaperones 106
Chesterton, G.K. 59
chirality problem 94
chlorophyll 88-89
chloroplasts 87
circumstellar habitable zone
    (CHZ) 61
Cleaver, Gerald 79
Clinton, Bill 103
codon 104-106
Collins, Francis 16
Collins, Robin 35

colostrum 128
Comet Shoemaker-Levy 9
Commentary on Genesis 122
communication 128, 143
compartmentalization 98
compatibilism 148
comprehensive philosophy 17
condensation 82
cones 131
consciousness 140-141
convection currents 65
convective zone 56
Copernican Principle 2-3
Copernicus, Nicolaus 2
Coriolis force 63
corotation circle 54
cosmic microwave background
    (CMB) 44-47
cosmological constant 41
    geography of 48-50
cosmological dark ages 46
Covalent bonds 84
Craig, William 33, 43
creatio ex nihilo 41
Cremonini, Cesare 3
Crick, Francis 18, 91, 95
cytokines 129
cytoplasm 105, 110
cytosine (c) 104
cytoskeleton 110

D

Darwin, Charles 113
Darwin's finches 119
Darwinism 113, 122
Davies, Paul
    on electromagnetism 36, 38
    on geography of universe 49
    on information 14, 99
    on multiverse 71
    on the Big Bang 42

Dawkins, Richard 79, 113
de Duve, Christian 92
deduction 6-7
Dembski, William 24, 109
dendrites 132, 134
Dennett, Daniel 140
Descartes, Rene 137
determinism 144
  and free will 148
  chemical determinism 117
  strict 145
deuterium 38, 46
Dingle, Herbert 24
Dirac, Paul 21
Divine creation 92
DNA 103-105, 116-117
doppler effect 42
double helix ladder 104
Dyson, Freeman 5
E. coli 118
Einstein, Albert
  on purposeful universe 5
  on science and God 14
  on laws of mathematics 24
  on modern quantum theory 148
electromagnetic force 36-38
electromagnetism 36, 38
electron 31-32
elements 31, 37
elliptical galaxy 52
Ellis, George 77
ENCODE (Enclyclopedia of DNA Elements) 108-109
endoplasmic reticulum 110
energy 37, 48, 99
  dark 41, 49
  fusion 51
  Hoyle 86
  renewable 58
entropy 43, 47-48
enzymes 97, 118

evolution 113, 115
  theistic (TE) 121
exponential numbers 25

F

fermions 32
Feynman, Richard 7
Fibonacci cascade 28
Fibonacci sequence 27-29
Fibonacci, Leonardo 27
fibrin 130-131
fibrinogen 131
Final Anthropic Principle (FAP) 5
fitness 114
fixation 90, 115
Flew, Anthony 111
Ford, Henry 13
free radicals 89
free will 144-147
  and compbatibilist option 148
functional magnetic resonance (fMRI) 139

G

galactic habitable zone (GHZ) 53
Galilei, Galileo 2, 21
gamma rays 36, 56
Garay, A. 94
gauge bosons 32
general theory of relativity 22, 41, 73
genes 104, 115
  junk 109
  hox 126
genome 103-105
Gingerich, Own 14
Gitt, Werner 100
glial cells 132
gluons 33, 37
God Particle 33-34

Godel, Kurt 77
God-of-the-gaps argument 12
golden spiral 27-28
golgi 110
Gonzalez, Guillermo 53
geocentric model 2-3
Gould, Stephen J. 13, 119
grand tack 69
gravity 35-36
   and Jupiter 69
   and stars 51-52
   and the Big Bang 41-42
   pressure 63
Greenstein, George 86
Gribbin, John 9, 85
Griffiths, Robert 17
Grossman, Lisa 83
guanine (G) 104

H

Hamilton, W.D. 127
haploid cell 127
Hartsfield, Tom 78
Hawking, Stephen
   on cause 43
   on free will 145
   on fundamental forces 34
Heile, Frank 76
heliocentric model 2-3
helper T-cells 129
Henderson, Lawrence 81
Higgs boson 32-34
Higgs field 32-33
Higgs, Peter 32
Hilbert space 72
histone 104
Hox genes 126
Hoyle, Fred
   and Big Bang 44
   on carbon 85-86
   on DNA/RNA 95, 111

   on evolutionary claims 113
Hsp90 (heat shock protein 90) 120
Hubble, Edwin 42
Human Genome Project (HGP)
   103-104, 108-109
hydrogen
   and human body 81-82
   and oxygen 83-84, 88
   and stars 51
   and sun 55
   helium/ 63
hydrologic cycle 82

I

immune system 128-129
induction 7
Infeld, Leopold 148
information 99-101
   and mind 139
   and necessity 115-117
Innanen, Kimmo 70
innate immune system 128-129
intelligent design 100, 116
International Journal of
   Cardiology 28
iron 52
   and cardiovascular system 130
   core 63, 68
irregular galaxy 52
isotopes 31, 37

J

Jastrow, Robert 8, 42, 44
Jesuits 3
Johnson, Phillip 119
Jupiter 68-70

K

Kalam argument 43

kenosis 123
Kenyon, Dean 91
Kepler, Johann 2
Kingsley, Charles 122
Koonin, Eugene 95

L

Lagrange, Joseph-Louis 12
language 137-138
   and conscious mind 142-143
Laplace, Pierre Simon 12-13
Large Hadron Collider (LHC) 33, 76
Laughlin, Robert 113
Lemaitre, Georges 14, 42
Lennox, John
   and abstract information 100
   and Ford example 13
   on abstract law and personal agency 75
   on God 34
lenticels 88
Leonard Award 9
lepton 32
Leslie, John 4
Level I 72-73
Level II 72-73
Level III 72-73
Level IV 73
Lewis, C.S. 59, 141
Lewontin, Richard 11
Libet, Benjamin 146
light year 53
Lightman, Alan 72
lightning 90, 93
Linde, Andrei 4
Lipton, Peter 7
lithosphere 64-65
Livio, Mario 26
luminosity 56-57
Lund University 119

lymphocytes 129
lysosomes 110

M

macroevolution 113
   and Cambrian explosion 119
   and time 117
macrophages 129
Maddox, John 43
magnetic field 63-64
magnetic shield 63-64, 76
Magnus, Albertus 15, 124
Manson, Neil 79
materialism 9, 91, 137
   and free will 144
   and naturalism 16-17
   Lewontin on 11
   methodological 17
   ontological 17, 44
   Penrose on 48
materialist science 91
mathematics 21-24
   of M theory 76
Mathematics of Evolution 113
Mayr, Ernst 121
McGilchrist, Iain 146
McGrath, Allister 111
McIntosh, Andrew 98
McLean, Euan 34
mediocrity principal 1
meiosis 127-128
Mendel, Gregor 14
Mercury 62
messenger RNA (mRNA) 100, 105
metabolism-first hypothesis 98-99
Meyer, Stephen 116
microevolution 113, 117, 120
Milky Way 51-54
Miller, Kenneth 123
Miller, Stanley 92
Miller-Urey experiment 92, 101

Millikan, Robert 3
misanthropic principle 1
mitochondria 111
mitosis 127-128
Mlodinow, Leonard 34
modal realism 73
monomers 93-94
moon 67-70
Mount Pinatubo 67
mountains 63, 65
M-theory 73
Muller, Gerd 117
multiverse 71-73
   and panspermia 95
   exists 78-79
Murray O'Hair, Madalyn 17
myelin 132
myelination 134

N

National Academy of Sciences (NAS) 119
National Aeronautics and Space Administration (NASA) 47
natural laws 16, 117
natural selection 113-114, 141
   and genetic mutations 118
   and necessity 115
naturalism 16-17
nebular cloud 55
necessity 115
neural plate 132
neuron 132-134
neurotransmitters 133
neutron 31-33
   and weak and strong forces 37-38
Nicolelis, Miguel 134
Nilsson, Nils 119
NIODA 6, 123

nitrates 90
nitrogen 89-90
NOMA (non-overlapping magisteria) 13
nuclear pore complex 105
nucleus 37-38, 110
Nuland, Sherwin 130
null hypothesis 7, 24

O

occipital lobe 131
optic nerve 131
orbital eccentricity 62
Orgel, Leslie 95
origin of life (OoL) 91-93, 99, 101
Orion Arm 54
oscilloscope 146
oxygen
   and cardiovascular system 130
   and photosynthesis 87-89
   as molecule of life 81-84
   atmosphere 92-93
ozone 64
ozone layer 64, 93

P

Page, Don 23
Pagels, Heinz 5
panspermia 95-96
participatory anthropic principle (PAP) 6
Penrose, Roger
   on brains and cosmos 132
   on mathematics 21, 26
   on multiverse notions 77
   on phase-space 47
Penzias, Arno 44
period of heavy bombardment 69
Perseus 54
phase-space 47

phenotype 114-115
phosphorus 51, 66
photoreceptors 131
photosynthesis
    and carbon-oxygen cycle 87-89
    and volcanos 66-67
Planck, Max 16, 18
planetesimals 83
Planinic, Josip 4
plasma 55, 130
plasma atmosphere 64
plasmin 131
plasminogen 131
plate tectonics 64-66
platelets 130
Politzer, Georges 43
Polkinghorne, John 123, 141
polymerization 93
polymers 93, 96
polymorphic gene 104
postmodernism 77
Pratt, Wallace 66
probability
    and free will 144
    and homochirlity 94-95
    boundary 25, 62
    limit of 24
    of Big Bang 47
Prost, Addy 99
proteins 93, 97
    folding 110
    making 104
Protestant Reformation 2
protons 31-32
proton-to-electron mass ratio 38
Ptolemy, Claudius 2
punctuated equilibrium theory 119

Q

quantum theory 33

quarks 32, 45
Quaternary stage 106
Quaternary Triplet Code 103
quartz 84
Quintus Tertullian 15

R

racemic 94
Radford, Tim 75
radiative zone 56, 63
radioisotopes 37
rationalism 6
reaction rates 92, 94
recombination era 45-46
red queen hypothesis 126-127
red-blood cells 130
Rees, Martin 37, 85
replication 95, 97
reproductive success 115, 141
resonance 85-86
respiration 88-89
ribosomal RNA (rRNA) 106
ribosomes 100, 110-111
ribozyme replicase 97
Ridley, Mark 127
Riemann, Bernhard 21
RNA 93
    RNA polymerase (RNAP) 105
    RNA-first hypothesis 97
    world hypothesis 96-98
rods 49
Ross, Hugh
    on exponential numbers 25
    on just-right tides 58
    on planet chances 9
    on solar system 53
Russell, Colin 11
Russell, Robert 14, 123

S

Sagan, Carl 61, 84
Sagittarius A* 53
Sandage, Allan 16, 44, 135
Saturn 62, 69
Schrodinger, Erwin 101
Scientific Dissent from Darwinism 113
secondary structure 106
Shalev, Baruch 17
Shostak, Seth 75
silicon 84
Silva, Ignacio 123
single-nucleotide polymorphism (SNP) 104
Sinha, S. 27
Sir Arthur Eddington 18, 22
Sir Isaac Newton 15
Sir James Jeans 18
Sir John Eccles 139
Smalley, Richard 78
soft determinism 145, 148
solar flares 63
speciation 119
Sperry, Roger 132
spin axis 68
spiral arms 53-54
spiral galaxy 52, 54
Spradley, Joseph 68
stagnant lid 65-66
standard model 31-32
Stark, Rodney 16
stars 51-52
   and gravity 35
   and strong and weak forces 37 38
   formation of 46
   in Orion Arm 54
   luminosity 56-57
stellar nucleosynthesis 46, 51, 85, 87

Steno, Nicolas 14
Stevens, John 131
stomata 87-88
Strauss, Michael 34
strict determinism 145
strings 74, 76
strong anthropic principle (SAP) 5
strong force 37
subduction 64
Sun 55-58
   and Earth's circling 63
   as metaphor for God 59
supernova 38, 46, 52
suppressor T-cells 129
Susskind, Leonard 50
Swindell, Rick 89
synapses 133-134
synaptogenesis 134

T

Taylor, Stuart 9
T-cells 129
Tegmark, Max 72
tertiary stage 106
tetrahedron 82
The Astonishing Hypothesis 18
The Fibonacci Quarterly 27
The Golden Ratio 25-26
   and Fibonacci sequence 27-29
The Grand Design 74
The Intelligent Universe 96
The Myth of Sisyphus 1
The Origin of Species 113
The Supernova Cosmology Project 40
theism 17
theories 7, 75-76
thermodynamics 47, 94
Thomson, Joseph J. 31
Through the Looking Glass 126
thymine (T) 104

Index   169

tidal locking 62, 69
tides 58-59
Tipler, Frank 5
transcription 105
transfer RNA (tRNA) 105
transitional forms 119
translation 95, 105-106
Treatise on Celestial Mechanics 12
triple alpha process 85

U

uncertainty principle 6
University of Bologna 15
University of California, Davis 57
uracil 105
Urey, Harold 92
UV radiation 64

V

vacuoles 110
valence electrons 84
Venus 63
vibration 85
virtual photons 36
Voyager 1 61

W

W boson 38
Wahlberg, Mats 122
Wald, George 92
Walker, Sara 99
water 81-84
    and RNA 96
    mass 63
Watson, Bruce 93
weak anthropic principle (WAP) 4
weak nuclear force 34, 38
Weber, Max 144
Wei-Haas, Maya 67

Weinberg, Steven 50, 79
Wheeler, John 3, 6
Wickramasinghe, Chandra 95-96
Wigner, Eugene 28, 140
Wilkinson Microwave Anisotropy
    Probe 48
Williams, George 115
Williams, Mark 83
wind 58, 68
Woit, Peter 77
Wood, Bernard 113

Y

Yockey, Hubert 100, 116

Z

Z boson 38
Zalasiewicz, Jan 83
zygote 126-128

www.ingramcontent.com/pod-product-compliance
Lightning Source LLC
Chambersburg PA
CBHW071400290426
44108CB00014B/1621